INSTRUCTOR'S MANUAL WITH TEST BANK
TO ACCOMPANY

EARTH SCIENCE
AND THE ENVIRONMENT

THOMPSON • TURK

GRAHAM R. THOMPSON
University of Montana
•
JONATHAN TURK
•
CHRISTINE SEASHORE

SAUNDERS COLLEGE PUBLISHING
Harcourt Brace Jovanovich College Publishers
Fort Worth Philadelphia San Diego New York Orlando Austin San Antonio
Toronto Montreal London Sydney Tokyo

Printed in the United States of America.

Thompson/Turk/Seashore: Instructor's Manual with Test Bank to accompany
EARTH SCIENCE AND THE ENVIRONMENT, 1/E

ISBN 0-03-029033-3

345 066 987654321

PREFACE

Earth Science encompasses geology, oceanography, meteorology, climatology, and astronomy: the sciences related to the Earth and other objects in the Universe. In Earth Science and the Environment we teach the fundamentals of each of these subjects and, at the same time, use their respective concepts to explain their roles in the human environment and the effects of human activities on the natural world.

We feel that Earth Science comes to life for both professionals and students when its concepts are used to explain specific and familiar events and phenomena. Because of this orientation, we use familiar examples and cite case histories throughout the book. We describe and explain basic concepts for each topic discussed in the text, and give them vitality and relevance by applying them immediately to specific examples. Whenever possible, we refer to places and events that are known to most students. In this way, we rely on and augment an existing familiarity and interest. This approach of applying general concepts to familiar places strongly reinforces the learning process and helps students maintain their interest in Earth Science long after the introductory course has ended.

Each of us who teaches Earth Science has developed his or her own ways of selecting, organizing, and presenting the topics of the course. Most of us do our best to capitalize on our own strengths. For these reasons, we do not believe that there is one best sequence for an introductory course, nor a particular way in which a topic should be presented. We do feel that this text presents topics in a logical order, and that information is presented so that it is useful as background for following chapters. However, other sequences are equally reasonable. Therefore, we have written the book so that individual chapters or groups of chapters can stand alone, and can be used in almost any sequence. Where information in a previous chapter is used in a later chapter, we refer the reader to the original discussion. In this way, if a student needs to have his or her memory refreshed, or wants a quick introduction to a concept, he or she can find it quickly.

Each chapter in this guide is divided into four sections.

(1) Discussion: In this section we highlight topics in the text that we feel are most important or interesting. In some cases we add information that is relevant to the chapter, but is not included in the text. Some professors may wish to incorporate this information into lectures or classroom discussions. In some cases, alternative teaching sequences are suggested where we feel that they may be helpful.

(2) Answers to discussion questions: The discussion questions in the text are designed to be thought-provoking and to serve as springboards for classroom discussion. Answers are given for some, and in instances where there are no "correct" answers, we discuss plausible responses based on the concepts presented in the chapter. Answers to the review questions can be gotten directly from the text of each chapter, so those answers are not repeated here.

(3) Bibliography: A selected bibliography is provided as reference for additional lecture material, or to provide references to students for further research into the topics.

(4) Test: A test consisting of multiple choice, true/false, and completion questions is provided for each chapter. An answer key follows each chapter test for the instructor's convenience.

Earth Science and the Environment

TABLE OF CONTENTS

CORRELATION CHART

Thompson: Instructor's Manual/Test Bank for EARTH SCIENCE AND THE ENVIRONMENT

In the computerized test bank every chapter is divided into three parts: multiple choice, true/false, and fill-in-the-blank. The multiple choice questions can be found in chapters followed by "M". The true/false questions can be found in chapters followed by "T". The fill-in-the-blank questions can be found in chapters followed by "F" (i.e.: Chapter 1M, Chapter 1T, and Chapter 1F).

CHAPTER 1

Earth Science and the Earth's Origin

Discussion

This chapter is an introduction to Earth Science, and thus we begin by describing Earth Science as the study of the "four realms of the Earth:" the atmosphere, the biosphere, the solid Earth, and the hydrosphere. In addition, Earth Scientists study extraterrestrial objects, both for their intrinsic interest, and because of information they provide about the origins of the Universe and, ultimately, of the Earth itself. These topics are treated by the sciences of geology (including paleontology - the history of life), oceanography, meteorology and climatology, and astronomy.

The bulk of the chapter is devoted to giving a brief, but informative, introduction to each of the subjects. Geology is described first, and the discussion includes brief overviews of rocks and minerals, the Earth's internal and surface processes, and geologic time. Introductions to the hydrosphere and oceanography, the atmosphere, astronomy, and the biosphere follow.

The final sections of the chapter introduce the origins of the Universe, Solar System, and Earth. This discussion leads easily into a description of the Earth's layers, and then to a brief introduction to plate tectonics in order to lay groundwork for following 3 chapters on minerals, rocks, and geologic time. We return to plate tectonics in Chapter 5 for a more complete discussion of the topic.

Chapter 1 includes short Focus On discussions of the nature of scientific theory, the big bang theory of the origin of the Universe, and evidence regarding the Earth's origin.

This brief introduction to the scope of Earth Science, and then to plate tectonics - one of the most fascinating aspects of geology - should spark the student's interest and impart a "feel" and excitement for Earth Science.

Answers to Discussion Questions

1. Most obviously, without water or an atmosphere, there would be no life. But, additionally, atmosphere and water are the primary agents of weathering and erosion; without them, the Earth would be more similar to the Moon. However, the Earth is larger than the Moon and has preserved more of its internal heat. Since tectonic processes reshape landforms, the Earth would still be differentiated into basins and continents, and the topography broken by mountains and mountain ranges.

2. Living organisms have interacted with geochemical processes to produce an atmosphere unique in the Solar System. Without life, the earth's atmosphere would be different.

3. Since water is essential for life, and the usable water is limited, excess consumption and pollution threaten our well-being.

4. By studying our neighbors in space we can learn about the evolution of our own planet. For example, ancient meteorite craters have been eradicated on Earth by tectonic and surface processes. But we can see them on the Moon, and since the Moon and Earth share a similar environment in space, we can deduce that the Earth was also bombarded by meteorites early in its history.

5. (a) During the evolution of the Solar System, most planetary matter was incorporated in the disk-shaped cloud that orbited the protosun. As the cloud coalesced, all the planets continued to orbit in the same direction as the original disk. A moving object will only change direction if acted upon by a force. Once the original dust cloud started to rotate, inertia kept the particles moving in the same direction. (b) See the answer to (a) above. In absence of an outside force, the newly coalesced planets would continue to orbit in the plane of the original disk. (c, d, and e) The dust cloud that evolved to form the Solar System was originally homogeneous. It was composed of about 75 percent (by mass) of hydrogen and 23 percent helium. The remaining 2 percent consisted of a mixture of all the naturally occurring elements. All of the elements are found on all the large objects in the solar system because the larger objects have strong gravitational fields. The compositional differences in the modern solar system are a result of escape and/or capture of matter. As explained in the text, the gravitational force of the Sun was so great that most of the hydrogen and helium and other light elements were retained. The four inner planets and their moons lost most of their light elements as they were "blown off" by solar winds, leaving rocky, metallic spheres of similar composition. Jupiter was massive enough and far enough from the Sun to retain most if its light elements. In fact, it may have grown by capture of gases that vaporized from the inner planets. Refer to Chapter 21 for further discussion.

6. If the longest half-lives of the Earth's radioactive isotopes were a few million years, the Earth would have become molten soon after its formation. Nearly all of these short half-lived radioactive isotopes would have decomposed during the 4.6 billion years of the Earth's history, so today our planet would be cooler than it is now. Volcanism, earthquake activity, and tectonic motion would be much less pronounced than it is today.

7. Heat escapes from a small sphere rapidly, for only a limited amount of rock exists to act as an insulator. The smaller planets and moons have cooled so much that they are no longer affected by volcanism, earthquake activity, and tectonic motion. (An exception is Io, which is heated by tidal forces as explained in Chapter 22.) In addition, small planets have insufficient gravitational fields to retain an atmosphere, and as a result, weathering is much slower. Thus, Mercury and the Moon, and to a smaller extent Mars, retain numerous meteorite craters formed 4 billion years ago, whereas similar ancient craters have been obliterated by weathering and tectonic activity on the larger planets, Earth and Venus.

Selected Bibliography

A readable history of geology:
Henry Faul and Carol Faul: It Began With A Stone. New York, John Wiley & Sons, 1983. 270 pp.

One of the best ways to teach geology is through field trips. Numerous guides to local areas are available, and many geology departments have their own guides or laboratory manuals containing guides. We recommend two excellent field guides, a series entitled Roadside Geology which has volumes for: Alaska, Arizona, Colorado, Montana, New Mexico, New York, Northern California, Oregon, Virginia, Washington, and the Yellowstone country. In addition, students who live in the northeast might enjoy:
Chet Raymo and Maureen E. Raymo: Written in Stone: A Geological History of the Northeastern United States. Chester, Connecticut, The Globe Pequot Press, 1989. 163 pp.

Two oceanography texts are:
Paul Pinet: Oceanography, An Introduction to the Planet Oceanus. St.Paul, MN, 1992. 551 pp.

Harold V. Thurman: Introductory Oceanography. 6th ed. New York, Macmillan Publishing, 1991. 514 pp.

In recent years there has been significant debate surrounding the questions: "What was the composition of the Earth's primitive atmosphere and how did life form?" Two references for further study are:

S.K. Atreya, J.B. Pollack, and M.S. Matthews, eds: Origin and Evolution of Planetary and Satellite Atmospheres. Tucson, University of Arizona Press, 1989. 881 pp.

Stephen Schneider and Randi Londer: The Coevolution of Climate and Life. San Francisco, Sierra Club Books, 1985. 576 pp.

Two books on weather systems weather are:

W.J. Burroughs: Watching the World's Weather. Cambridge, Cambridge University Press, 1991. 196 pp.

Leslie Musk: Weather Systems. Cambridge, Cambridge University Press, 1988. 158 pp.

A general textbook on climate is:
Michael R. Rampino, John E. Sanders, Walter S. Newman, and L.K. Konigsson, eds: Climate. New York, Van Nostrand Reinhold, 1987. 608 pp.

A more in depth coverage of the formation of the Universe is given in:
Steven Weinberg: The First Three Minutes: A Modern View of The Origin of the Universe. New York, Bantam Books, 1977. 169 pp.

John Gribbin: Genesis: The Origins of Man and the Universe. New York, Delacorte Press/Eleanor Friede, 1981. 360 pp.

For additional reference on the planets we recommend two standard astronomy texts (Also see additional references given in Chapter 22.):

George O. Abell, David Morrison, and Sydney C. Wolff: Realm of the Universe Fourth Edition. Philadelphia, Saunders College Publishing, 1988. 529 pp.

Jay M. Pasachoff: Contemporary Astronomy Fourth Edition. Philadelphia, Saunders College Publishing, 1989. 579 pp.

The formation of the Earth is discussed in:

George W. Wetherill: Formation of the Earth. Annual Review of Earth and Planetary Sciences 18 (205), 1990.

Chapter 1

Earth Science and the Earth's Origin

Multiple Choice:

1. The gaseous outer layer of our planet is the
(a) atmosphere; (b) biosphere; (c) hydrosphere; (d) solid Earth.

2. The hydrosphere is composed mainly of
(a) rock; (b) soil; (c) water; (d) gases.

3. An example of an internal process is a
(a) landslide; (b) volcanic eruption; (c) erosion of a river valley; (d) deposition of rock by an avalanche; (e) deposition of rock by a glacier.

4. An example of a surface process is
(a) weathering; (b) a volcanic eruption; (c) an earthquake; (d) a geyser; (e) formation of a cavern by ground water.

5. Most fresh water on Earth is found in
(a) oceans; (b) rivers; (c) glaciers; (d) ground water; (e) lakes.

6. The Earth and planets all formed from a single cloud; today
(a) no two planets are alike; (b) they remain unchanged; (c) they are mostly alike; (d) they all have an atmosphere similar to Earth's.

7. The age of the Universe starting with the Big Bang is about
(a) 800,000 to 1 million years; (b) 8 to 20 billion years; (c) 20 to 40 trillion years; (d) 10 to 30 million years.

8. The heat and light given off by our modern Sun is generated by
(a) meteoroids; (b) burning hydrogen; (c) gravitational coalescence; (d) hydrogen fusion.

9. The Earth is _____ years old.
(a) 6000; (b) 100 million; (c) 4.6 million; (d) 4.6 billion; (e) 10 to 20 billion.

10. Mercury, Venus, Earth, and Mars are called the
(a) Plutonic planets; (b) metallic planets; (c) terrestrial planets; (d) Jovian planets.

11. The crust is ______ thick.
(a) 7 to 70 kilometers; (b) 600 to 6000 kilometers; (c) 1.1 to 11 millimeters; (d) 10 to 20 meters

12. The ____ of the Earth contains over 80 % of its volume.
(a) core; (b) mantle; (c) crust; (d) asthenosphere

13. Which part of the Earth is composed mainly of liquid iron and nickel?
a) crust; (b) core; (c) mantle; (d) asthenosphere

14. The lithosphere consists of
(a) the crust; (b) the mantle; (c) the lower mantle and upper crust; (d) the crust and upper mantle.

15. Which term means "weak layer"?
(a) lithosphere; (b) asthenosphere; (c) hydrosphere; (d) biosphere; (e) atmosphere

16. According to plate tectonic theory, the lithosphere is segmented into __ major plates and several smaller ones.
(a) 7; (b) 13; (c) 20; (d) 3

17. A zone where 2 or 3 plates separate or move apart is called
(a) a convergent boundary; (b) a transform boundary; (c) subduction zone; (d) a divergent boundary.

18. A submarine mountain range running north-south through the Atlantic Ocean is
(a) a subduction zone; (b) the Mid-Atlantic ridge; (c) a transform boundary; (d) a divergent boundary.

19. Volcanoes and earthquakes are common at
(a) subduction zones; (b) rift valleys; (c) mid-oceanic ridges; (d) all of the above.

20. The San Andreas fault in California is an example of
(a) a divergent boundary; (b) a transform boundary; (c) a convergent boundary; (d) a subduction zone

True or False:

1. The hydrosphere includes water in streams, lakes, and oceans; in the atmosphere; and frozen in glaciers.

2. The atmosphere is a mixture of gases, mostly nitrogen and hydrogen.

3. The biosphere is the thin zone inhabited by life.

4. Below a thin layer of soil and beneath the ocean water, the outer layers of the Earth are composed entirely of molten rock.

5. Biologists study rocks, their composition, their formation, and their behavior.

6. Earthquakes and volcanic eruptions are internal processes.

7. Stream erosion that forms a canyon is an example of an internal processes.

8. The Earth is about 4.6 billion years old.

9. Oceans cover 71 percent of the Earth and contain 97.5 percent of its water.

10. Enough ice remains in Antarctic and Greenland icecaps to raise sea level by 65 meters if it were to melt completely.

11. Although streams and lakes contain most of our visible fresh water, 70 times more water is stored underground, as ground water.

12. The Earth's atmosphere acts as a blanket, retaining heat at night and shielding us from direct solar heating during the day.

13. Mercury, the planet closest to the Sun, has an atmosphere similar to Earth's.

14. Human activity is significant enough to affect most internal geological processes such as volcanic eruptions, earthquakes, and the movement of continents.

15. Most astronomers place the start of time and the origin of the Universe between 15 and 18 billion years ago.

Completion:

1. The ______ includes water in streams, lakes, and oceans; in the atmosphere; and frozen in glaciers. It also includes ground water that soaks soil and rock to a depth of 2 or 3 kilometers.

2. The ______ is a mixture of gases, mostly nitrogen and oxygen.

3. ______ is the study of the solid Earth: its rocks and minerals, the physical and chemical changes that occur on its surface and in its interior, and the history of the planet and its life.

4. Events and processes that occur or originate within the Earth are called ______ ______.

5. ______ is the study of all the Earth's water, its distribution, and its circulation among oceans, continents, and the atmosphere.

6. ______ is the study of the world's oceans, including coastlines, topography of the sea floor, the nature of seawater, ocean currents, and marine life.

7. ______ is the state of the atmosphere at a particular time and place.

8. ______ is the characteristic weather of a region over long periods of time.

9. The ______ is the thin layer at the Earth's surface containing all life.

10. The ______ is the cataclysmic event that marked the beginning of the Universe and the start of time.

11. The earliest form of the Sun is the ______.

12. The ______ of the Earth is a thin, rigid surface veneer ranging from 7 kilometers under some portions of the oceans to a maximum of about 70 kilometers under the highest mountain ranges.

13. The thick, mostly solid layer called the ______ lies beneath the crust and surrounds the core.

14. The cool, strong, brittle outer portion of the Earth, including both the crust and the upper layer of mantle, is called the ______.

15. A ______ ______ is a zone where 2 or sometimes 3 plates separate, or move apart from each other.

Answers for Chapter 1

Multiple Choice: 1.c; 2.c; 3.b; 4.a; 5.c; 6.a; 7.b; 8.d; 9.d; 10.c; 11.a; 12.b; 13.b; 14.d; 15.c; 16.a; 17.d; 18.b; 19.a; 20.b

True or False: 1.T; 2.F; 3.T; 4.F; 5.F; 6.T; 7.F; 8.T; 9.T; 10.T; 11.T; 12.T; 13.F; 14.F; 15.T

Completion: 1. hydrosphere; 2. atmosphere; 3. geology; 4. internal processes; 5. hydrology; 6. oceanography; 7. weather; 8. climate; 9. biosphere; 10. Big Bang; 11. protosun; 12. crust; 13. mantle; 14. lithosphere; 15. divergent boundary

CHAPTER 2

Minerals

Discussion

Chapters 2 and 3 discuss minerals and rocks, the materials that make up the Earth. We discuss both the nature of minerals and rocks and the processes that form and change them.

At least two aspects of minerals commonly fascinate introductory geology students because they are familiar: crystals and gems. A professor can stimulate class interest by pointing out that the striking faces on diamonds and other gems results from the perfect ordering of atoms in the crystals, and that many gems are simply beautiful varieties of common minerals.

The knowledge that only a few minerals are common, and that a student can learn to identify and name them easily with a bit of practice in lab, is another aspect of the study of minerals that usually draws students into the subject. It seems to give students a good feeling to know that only a few minerals are abundant, and therefore they can easily learn to identify most minerals in most rocks.

In Chapter 2 we begin by pointing out that the Earth's continents are made of solid rock. In turn, all rocks are made of one or more minerals. Therefore, minerals are the fundamental building blocks of the Earth. For this reason, an appreciation of the nature of minerals and a basic knowledge of the few, common, rock-forming minerals lays a foundation for the study of the Earth.

The chapter then explains that the essential nature of a mineral is that it has a crystalline structure -- an orderly, repetitive, spatial organization of atoms - and a definite chemical composition. Because chemical composition and chemical bonding are important to mineralogy, we provide a succinct discussion of the chemical composition of the Earth's crust, and of chemical bonding as it applies to minerals. This discussion is a simple and adequate introduction to the subject for students with no chemistry background, and will be serve as a quick review for students who have taken introductory chemistry.

We then describe the physical properties of minerals, and explain how they can be used to identify minerals. We stress the point, however, that most field identification of minerals is actually done by recognition of common minerals, aided by a few simple tests of physical properties. The 9 rock-forming minerals are then described, along with a few interesting ore minerals, gems, and accessory minerals. Those 20 or 30 minerals are the most useful to be able to identify.

The chapter includes an Earth Science and the Environment discussion of asbestos and its relationships to cancer and other diseases.

Answers to Discussion Questions

1. Two properties differentiate one mineral from all others: chemical composition and crystal structure. If diamond and graphite are different minerals, but have identical chemical compositions, they must have different crystal structures. Diamond is cubic, and all carbon-carbon bonds are strong covalent bonds. Graphite is hexagonal, and each carbon is bonded to three other carbons by similar covalent bonds, but to a forth carbon with a weak van der Waals bond. It is this weak bond that is responsible for the softness and cleavage of graphite, and for its properties of electrical conductance and opacity as well.

2. Silicon and oxygen combine to form the silica tetrahedron. Since all of the Earth's silicon, and most of its oxygen, combine in this way, the silica tetrahedron accounts for nearly 75 weight percent of the Earth's crust. But the silica tetrahedron has a minus 4 charge; it is a complex anion. Its charge must be satisfied by sharing oxygens and the bonding of additional cations to the silica tetrahedron. Those cations make up the difference between 75 and 95 percent.

3. The minerals on the Moon, Mars, and Venus should be similar to those found on Earth. The compositions of the Moon, Mars, and Venus are similar to that of Earth, and the laws of physics and chemistry are identical everywhere. Therefore, the minerals should be identical.

Selected Bibliography

This or any one of a number of other textbooks used for teaching introductory mineralogy, provides detailed background for the study of minerals, and most include an introduction to petrology as well:
Cornelius Klein and Cornelius Hurlbut: Manual of Mineralogy (after James D. Dana), 20th Ed. New York, John Wiley & Sons, 1985. 596 pp.

Nice color photos of minerals:
Iris Vanders and Paul Kerr: Mineral Recognition. New York, John Wiley & Sons. 1967.

Two good paperback guides to mineral and rock identification, with color photos are:
C. Chesterman: The Audubon Society Field Guide to North American Rocks and Minerals. New York, Alfred A. Knopf, 1978.

M. Prinz, G. Harlow, and J. Peters: Simon & Schuster's Guide to Rocks and Minerals. New York, Simon & Schuster, 1978.

The asbestos conflict is discussed in:
H. Schreier: Asbestos in the Natural Environment. Amsterdam, Elsevier Publishers, 1989. 172 pp.

Chapter 2

Minerals

Completion:

1. A naturally occurring, inorganic solid with a definite chemical composition and a crystalline structure is
(a) a rock; (b) an ore; (c) an element; (d) a mineral

2. Which of the following best describes an element?
(a) a fundamental form of matter that cannot be broken into simpler substances by ordinary chemical processes; (b) a fundamental form of matter that can be broken into simpler substances by ordinary chemical processes; (c) a naturally occurring, inorganic solid with a definite chemical composition and a crystalline structure; (d) a small, dense, positively charged center surrounded by a cloud of negatively charged electrons

3. How many chemical elements occur naturally in the Earth's crust?
(a) 8; (b) 10; (c) 27; (d) 88; (e) 108

4. The nucleus of an atom contains
(a) electrons; (b) ions; (c) shells; (d) protons; (e) elements

5. In a neutral atom the number of protons
(a) is greater than the number of electrons; (b) is less than the number of electrons; (c) is twice the number of electrons; (d) equals the number of electrons

6. Which has mass but no charge?
(a) proton; (b) electron; (c) ion; (d) neutron; (e) anion

7. An ion has
(a) an equal number of positive and negative charges ; (b) a net positive or negative charge; (c) mass but no charge; (d) an equal number of protons and electrons.

8. A cation is
(a) a negatively charged ion; (b) a positively charged ion; (c) a neutral ion; (d) a type of electron.

9. Ionic bonds, covalent bonds, metallic bonds, and van der Waals forces are all examples of
(a) chemical bonds; (b) cleavage; (c) luster; (d) streak.

10. When cations and anions bond together to form a mineral,
(a) the negative charges are greater than the positive ones; (b) the negative charges exactly equal the positive ones; (c) the

negative charges are less than the positive ones; (d) none of the above.

11. In which kind of bond do atoms mutually share electrons?
(a) covalent; (b) ionic; (c) van der Waals; (d) anionic; (e) noble

12. Metals are excellent conductors of electricity and heat because
(a) they have covalent bonds; (b) the electrons are not free to move; (c) the electrons are free to move, (d) they are not dense.

13. Which of the following does not describe a crystal?
(a) halite; (b) a substance with an irregular, unorderly pattern; (c) orderly rows and columns of alternating sodium and chloride ions; (d) any substance whose atoms are arranged in a regular, orderly, periodically repeated pattern.

14. Calcite is a/an
(a) silicate; (b) carbonate; (c) clay mineral; (d) element

15. Which of the following is not a rock-forming mineral?
(a) feldspar; (b) galena; (c) quartz; (d) olivine

16. Which physical properties distinguish each mineral from all others?
(a) streak and luster; (b) fracture and faceting; (c) van der Waals bonds; (d) chemical composition and crystal structure

17. The tendency of a mineral to break along flat surfaces is
(a) fracture; (b) luster; (c) faceting; (d) cleavage.

18. Mohs' hardness scale for minerals
(a) refers to the resistance of a mineral to fracture; (b) tells which minerals shatter most easily; (c) indicates how bendable a mineral is; (d) refers to the resistance of a mineral's surface to scratching.

19. Silicates, carbonates, and oxides are classified
(a) according to their cations; (b) according to their protons; (c) according to their anions; (d) according to their neutrons.

20. All minerals containing silicon and oxygen are
(a) glassy; (b) carbonates; (c) silicates; (d) quartz; (e) silicon.

21. Which mineral makes up more than 50 percent of the Earth's crust and is the most abundant mineral?
(a) mica; (b) feldspar; (c) quartz; (d) pyroxene; (e) amphibole.

22. A mineral that is commercially valuable for its beauty rather than for industrial use is

(a) a gem; (b) an ore; (c) halite; (d) olivine.

23. A silica tetrahedra is
(a) a pyramid with 4 silicon atoms at the corners and an oxygen atom in its center; (b) a pyramid with 4 oxygen atoms at the corners and a silicon atom in its center; (c) a ring of alternating silicon and oxygen atoms; (d) a chain of alternating silicon and oxygen atoms.

True or False:

1. It is not necessary for a substance to be a solid to be a mineral.

2. An element can be broken into smaller substances by ordinary chemical processes.

3. Only eight elements -oxygen, silicon, aluminum, iron, calcium, magnesium, potassium, and sodium- make up more than 98 percent of the Earth's crust.

4. The nucleus is made up of neutrons and electrons.

5. Atoms with a positive or negative charge are called ions.

6. A covalent bond forms when nearby atoms capture other electrons.

7. A crystal is any substance in which atoms are arranged without a regular, orderly, periodically repeated pattern.

8. Fracture is the way in which a mineral breaks other than along planes of cleavage.

9. The 9 rock-forming minerals are the least abundant minerals in rocks.

10. Quartz is the only silicate mineral that contains no cations other than silicon.

11. Individual clay crystals are large enough to be seen with the naked eye.

12. An ore is a mineral that contains commercially valuable elements or compounds.

Completion:

1. A/an ______ is a naturally occurring, inorganic solid with a definite chemical composition and a crystalline structure.

2. Electrical forces that hold atoms together into compounds are _____ _____.

3. Weak electrical forces that loosely bond atoms are _____ _____ _____ forces.

4. The tendency of a mineral to break along certain atomic planes is _____.

5. The outer, planar surface of a crystal that grows in an unimpeded manner is a _____ _____.

6. A scale that compares the resistance of the surfaces of minerals to scratching is _____ _____ _____.

7. ______ is the color of a fine powder of a mineral.

8. Every silicon atom surrounds itself with ______ oxygens in a silica tetrahedron.

9. In ______ ______ silicates adjacent tetrahedra do not share oxygens.

10. The ______ ______ consist of two single chains cross-linked by the sharing of more oxygens.

11. ______ is the most common type of potassium feldspar.

12. ______ is widespread and abundant in continental rocks, but rare in oceanic crust and the mantle.

13. Two carbonate minerals that are common in near-surface rocks of the continents are ______ and _______.

14. Two ore minerals that are comprised of pure metals are native _____ and ______.

15. Most ______ rocks start out as shell fragments and other hard parts of marine organisms.

Answers for Chapter 2

Multiple Choice: 1. d; 2. a; 3. d; 4. a; 5. d; 6. d; 7. b; 8. b; 9. a; 10. b; 11. a; 12. c; 13. b; 14. b; 15. b; 16. d; 17. d; 18. d; 19. c; 20. c; 21. b; 22. a; 23. b

True or False: 1. F; 2. F; 3. T; 4. F; 5. T; 6. F; 7. F; 8. T; 9. F; 10. T; 11. F; 12. T

Completion: 1. mineral; 2. chemical bonds; 3. van der Waals; 4.cleavage; 5. crystal face; 6. Moh's hardness scale; 7. streak; 8. four; 9. independent tetrahedra; 10. double-chain silicates; 11. orthoclase; 12. quartz; 13. calcite and dolomite; 14. native gold and silver; 15. carbonate

CHAPTER 3

Rocks

Discussion

We begin this chapter by explaining that an understanding of rocks and how they form allows an Earth Scientist to interpret much about the geologic history of that part of the Earth where the rocks occur. The chapter then describes the rock cycle as a paradigm of the fact that, in geologic time, rocks constantly change from one to another of the three main categories of rocks: igneous, sedimentary, and metamorphic.

The rest of the chapter explains the processes by which igneous, sedimentary, and metamorphic rocks form, describes how common rocks of each type are classified and named, and discusses the most common kinds of rocks in the Earth's crust.

In an introductory Earth Science text, the subject of rocks is one that can cover several chapters, and go on at great length because of the nearly infinite variety of rock types and processes by which they form. We have chosen to cover igneous, sedimentary, and metamorphic rocks in a single chapter by taking advantage of the fact that only a few kinds of rocks are common and encountered frequently by most of us, students and professionals alike. Those few, abundant rock types deserve to be stressed, and the myriad others can be ignored in an introductory course. Among the igneous rocks, we stress granite and basalt, and describe briefly rhyolite, gabbro, andesite, diorite, and peridotite (because it is the material of the mantle). We focus on shale, sandstone, and limestone among the sedimentary rocks, and slate, schist, gneiss, and marble as metamorphic rocks. We have chosen this approach to stressing the few, common rock types because we feel that it is more meaningful to an introductory student to learn to recognize common rocks, and understand the processes that form them, than to be impressed (or depressed) by the great number of uncommon rocks that have been named.

This chapter contains boxes describing the origins and effects of radon gas, and on Bowen's reaction series.

Answers to Discussion Questions

1. Magma begins to rise as soon as it forms because it is substantially (about 10 percent, on the average) less dense than the rocks surrounding it, and, being liquid, is mobile.

2. In the San Juans, rising granitic and intermediate magma erupted as ash flow tuffs and related volcanic rocks. As the magma rose, portions solidified at shallow depth in the crust. Thus, both intrusive and extrusive rocks formed from the same magma, or from closely related magmas.

3. Texture is the simplest clue to volcanic or plutonic origin of a rock. If an igneous rock is uniformly medium or coarse grained, it is probably plutonic. If it is glassy or very fine grained, it is probably volcanic. However, texture is not always reliable as an indicator of volcanic or plutonic origin. A thin dike or sill (plutonic) may have a fine texture if it was intruded into cool country rock. Some granitic volcanics are medium, or even coarse, grained, particularly if they crystallized at depth to a considerable extent before rising and flowing onto the surface. Thus, it is not always possible to distinguish volcanic or plutonic origin from a hand sample of a rock. It is much better to be able to study field relationships.

4. This question should have a simple answer, but it does not, partly because of the great variety of rock types in each of the three main categories. For most rocks, the following criteria work to identify them as igneous, sedimentary, or metamorphic.

Most igneous rocks are silicates, contain predominantly feldspar, show igneous textures (of which there is a wide variety), and contain characteristic mineral assemblages summarized in Figure 3-7 of the text. In the field, they show intrusive or extrusive relationships with country rock.

Most sedimentary rocks show bedding or other sedimentary structures, and most clastic rocks show rounding of the grains. Feldspar is typically much less abundant in sedimentary rocks than in igneous rocks, because it weathers much more easily than quartz or clay minerals, which are the most common minerals in clastic sedimentary rocks. Many sedimentary rocks contain fossils. Limestone is predominantly calcite (or dolomite), and has an unmistakable appearance to an experienced field geologist. This latter observation frustrates rather that helps introductory students, however.

Many metamorphic rocks show metamorphic foliation, but it is sometimes difficult to explain to an introductory student how to distinguish metamorphic layering from sedimentary layering. But metamorphic rocks usually also show interlocking grain boundaries rather than the rounded grains characteristic of clastic sedimentary rocks. Most metamorphic rocks contain minerals characteristic of metamorphism, such as chlorite, micas, garnet, and a variety of amphiboles. However, many of those minerals are not really unique to metamorphic rocks.

An additional difficulty in distinguishing among the three rock categories arises with rocks that fall on a boundary between two categories. An example of this type of rock is an air-fall tuff. It is both igneous and sedimentary. Similarly, a basal surge deposit of an ash flow tuff from a violent caldera eruption may show beautiful cross bedding. Another example of a boundary transcendent rock is the metamorphosed granite that comprises most continental basement rock.

This question is an excellent one to get students thinking about identifying rocks, and about how they form.

5. Sediment originates by weathering. Feldspar is the most abundant mineral in the Earth's crust. Feldspar (and other aluminosilicates) form mostly clay minerals when they weather. Therefore, clays are the most abundant products of weathering of the Earth's crust. When clays are deposited and lithified,

they form shale. Since clay is the most abundant weathering product, shale is the most common sedimentary rock.

6. Since feldspar weathers much more readily than quartz, the presence of so much feldspar in a sandstone implies that the sediment did not weather for long before it was deposited and lithified. Many rocks of this type formed close to the source of the sediment.

7. Sedimentary clasts become rounded during transport by water or wind. The rounding occurs as the clasts bounce together, or against bedrock during transport. Well-rounded grains imply a relatively long time and distance of transport, whereas angular grains imply less time and distance. Large clasts such as cobbles and boulders, however, commonly become well-rounded within short times and distances from their source. Sand and silt become rounded more slowly.

Angular clasts may also imply that water or wind was not the transport medium. They are common in debris flows and glacial deposits, where the viscosity of the medium cushions the grain impacts and slows rounding.

8. Contact metamorphic rocks commonly form a relatively small halo around the intrusion that caused the metamorphism, with metamorphic grade increasing toward the contact. Regional metamorphic rocks normally occur over a much larger area, and the metamorphic grade may bear little or no relation to plutons. Contact metamorphic rocks normally show no foliation, whereas most regionally metamorphic rocks are foliated.

9. The original sedimentary layering may be preserved as compositional layering in the metamorphic rock. Thus a metamorphosed sandstone layer will have a different composition from that of an adjacent metamorphosed shale layer. However, bedding within a single sedimentary rock type is often completely lost as slaty cleavage, schistosity, or gneissic textures develop.

10. If country rock similar to granite in composition (this includes most of the continental crust) is heated to the highest grades of metamorphism, the rock may partially melt to form granitic magma. That magma may remain in small veins in the country rock to form migmatite, or it may coalesce and rise through the crust to form a pluton.

Selected Bibliography

The most commonly used English-language text on igneous and metamorphic rocks is also an excellent literature review and reference:

Donald Hyndman: Petrology of Igneous and Metamorphic Rocks, 2nd Ed.. New York, McGraw-Hill Book Company, 1985.

An introductory text which is less detailed and simpler than Hyndman, but informative and useful for introductory geology students who may wish to read a bit more about igneous rocks is:
R. Dietrich and B. Skinner: Rocks and Rock Minerals. New York, John Wiley & Sons, 1979.

Sedimentary rocks are discussed in:
H. Blatt: Sedimentary Petrology. New York, W. H. Freeman, 1982. 564 pp.

H. Blatt, G. Middleton, and R. Murray: Origin of Sedimentary Rocks, 2nd Ed.. Englewood Cliffs, N. J., Prentice-Hall, 1980.

F. Pettijohn: Sedimentary Rocks, 3rd Ed.. New York, Harper & Row, 1975.

H. Reineck and I. Singh: Depositional Sedimentary Environments, 2nd Ed.. New York, Springer-Verlag, 1980. 439 pp.

Metamorphic rocks are discussed in:
F. Turner: Metamorphic Petrology, 2nd Ed.. New York, McGraw-Hill, 1980.

An excellent all-around reference book for Earth scientists is:
K. Krauskopf: Introduction to Geochemistry. New York, McGraw-Hill, 1967.

Chapter 3

Rocks

Multiple Choice:

1. Igneous rocks are those that form by
(a) lithification of sediment; (b) textural or mineralogical alteration of existing rocks; (c) solidification of magma; (d) precipitation of seawater.

2. Rocks of all kinds decompose or weather into sediment which with time cements together to form
(a) metamorphic rock; (b) igneous rock; (c) magma; (d) sedimentary rock.

3. Sedimentary rock makes up
(a) less than 5 percent of the Earth's mantle; (b) more than 5 percent of the Earth's crust; (c) 50 percent of the Earth's crust; (d) 12 percent of the Earth's core.

4. Metamorphic rock forms when
(a) magma cools and solidifies; (b) igneous, sedimentary, or other metamorphic rocks change because of high temperature and/or pressure or are deformed during mountain building; (c) seawater precipitates; (d) sediments are lithified.

5. The rock cycle refers to the fact that
(a) rocks remain the same over time; (b) igneous, sedimentary, and metamorphic rocks do not change over time; (c) rocks continually transform from one type to another over time; (d) sedimentary rocks change to igneous rocks.

6. Intrusive igneous rocks form when magma solidifies
(a) on top of the Earth; (b) within the Earth, before it can rise all the way to the surface; (c) in the core; (d) on the sea floor.

7. Extrusive igneous rocks form when
(a) magma solidifies within the Earth; (b) magma solidifies within the core; (c) sediment cements together; (d) magma erupts and solidifies on the Earth's surface.

8. The most abundant igneous rock in continental crust is
(a) basalt; (b) granite; (c) rhyolite; (d) porphyry: (e) gabbro.

9. ______ has the same mineral content as granite?
(a) basalt; (b) quartz; (c) rhyolite; (d) carbonatite: (e) gabbro

10. Gabbro is abundant
(a) at the Earth's surface; (b) in deeper parts of oceanic crust; (c) in sedimentary basins; (d) on the continents.

11. Clastic sediment describes
(a) fragments of weathered rock or of shells and other organic remains; (b) dissolved ions; (c) precipitated minerals; (d) clay particles.

12. The conversion of loose sediment to hard rock is called
(a) compaction; (b) precipitation from seawater; (c) lithification; (d) metamorphism.

13. Lithified sand grains form
(a) conglomerate; (b) granite; (c) sandstone; (d) shale.

14. The layering that develops as sediment is deposited is called
(a) bedding or stratification; (b) the dolomite problem; (c) ripple marks; (d) segmented layering

15. The process by which rocks and minerals change because of changes in temperature, pressure, or other environmental conditions is called
(a) lithification; (b) metamorphism ; (c) cementation; (d) stratification .

16. Rocks bend or break in response to the movement of tectonic plates, this change is called
(a) sedimentation; (b) extrusion; (c) deformation; (d) lithification.

17. Contact metamorphism refers to metamorphism
(a) that affects broad regions of the crust; (b) caused by intrusion of cold magma into hot rocks; (c) caused by the intrusion of hot magma into cooler rocks; (d) that is usually accompanied by deformation.

18. Most slate is formed from metamorphism of
(a) granite; (b) conglomerate; (c) slate; (d) rhyolite.

19. The changes in rock caused by migrating hot water and by ions dissolved in the hot water is called
(a) hydrothermal metamorphism; (b) cementation; (c) lithification; (d) skarn production.

True or False:

1. The most common sedimentary rocks are shale, sandstone, and limestone.

2. In the upper mantle, between depths of 100 and 350 kilometers, the temperature is so high that in certain places large amounts of rock melt to form magma.

3. Liquid magma is denser than its crystalline equivalent.

4. Rocks with especially high magnesium and iron concentrations are called ultramafic.

5. Gabbro is mineralogically identical to basalt.

6. In certain environments dissolved ions may precipitate directly to form clastic sediment.

7. Calcite, quartz, and iron oxides are the most common cements in sedimentary rocks.

8. Deposition of clastic sediment occurs when transport stops, usually because the wind or water slows down and loses energy or, in the case of glaciers, when the ice melts.

9. Evaporites are metamorphic rocks formed when evaporation of water concentrates dissolved ions to the point where they precipitate from solution.

10. Cross-bedding is an arrangement of small beds lying at an angle to the main sedimentary layering.

11. Graded bedding is a type of bedding in which the largest grains usually collect at the top of a layer and the grain size decreases toward the bottom.

12. Marble is made of calcite.

13. Foliation is not a kind of metamorphic layering.

14. Contact metamorphism of limestone forms tactite.

15. Regional metamorphism is usually accompanied by deformation, so the rocks are foliated.

16. Low-grade metamorphism occurs at shallow depths, more than 10 kilometers beneath the surface, where temperature is higher than 300° to $400^{o}C$.

17. At high metamorphic grades, light- and dark-colored minerals often separate into bands a centimeter or more thick, to form a rock called schist.

18. Most hydrothermal alteration is caused by circulating ground water, water contained in soil and bedrock.

Completion:

1. Geologists separate rocks into three classes based on how they form: ______, ______, and ______ rocks.

2. Igneous rocks that erupt and solidify at the earth's surface are _____.

3. Large crystals in a porphyry are _____.

4. The size, shape and arrangement of igneous mineral grains in a rock define its _____.

5. If volcanic magma solidifies within a few hours of erupting, volcanic glass, called ______ may form.

6. Igneous rocks with abundant silicon and aluminum are _____.

7. Igneous rocks with abundant magnesium and iron are _____.

8. The volcanic rock most common in the western mountains of South America is _____.

9. ______ is the most common rock in continental crust.

10. ______ is made of the same minerals as granite, but has a fine-grained texture because it is volcanic.

11. Although rare in the Earth's crust, ______ is an ultramafic igneous rock that makes up most of the upper mantle.

12. All solid particles transported and deposited by water, wind, glaciers, and gravity are referred to as ______.

13. Rocks that are composed of fragments of preexisting rocks and other particles that have been physically transported and deposited are called ______ ______ ______.

14. Thin bedding along which rock easily splits is called ______.

15. As peat is buried and compacted by overlying sediment, it converts to ______.

16. Calcite-rich carbonate rocks are called ______.

17. ______ is bioclastic limestone consisting wholly of coarse shell fragments cemented together.

18. Remains or traces of a plant or animal preserved in rock are called ______.

19. When rocks are buried, they become ______ and the pressure on them increases.

20. ______ is the change in shape of rocks in response to tectonic forces.

21. The ______ ______ of a rock is the intensity of metamorphism that formed the rock.

Answers for Chapter 3

Multiple Choice: 1. c; 2. d; 3. a; 4.b; 5. c; 6. b; 7. d; 8. b; 9. c; 10. b; 11. a; 12. c; 13. c; 14. a; 15. b; 16. c; 17. c; 18.c; 19. a

True or False: 1. T; 2. T; 3. F; 4. T; 5. T; 6. F; 7. T; 8. T; 9. F; 10. T; 11. F; 12. T; 13. F; 14. T; 15. T; 16. F; 17. F; 18. T

Completion: 1. igneous, sedimentary, and metamorphic; 2. volcanic or extrusive; 3. phenocrysts; 4. texture; 5. obsidian; 6. sialic; 7. mafic; 8. andesite; 9. granite; 10. rhyolite; 11. peridotite; 12. sediment; 13. clastic sedimentary rocks; 14. fissility; 15. coal; 16. limestone; 17. coquina; 18. fossils; 19. hotter; 20. deformation; 21. metamorphic grade

CHAPTER 4

Geologic Time, Fossils, Evolution, and Extinction

Discussion

The first objective of this chapter is to explain how Earth scientists measure the ages of rocks that formed, and events that occurred, in prehistoric time. The second objective is to give students a feel for the great length of geologic time, and to make it clear that events that are improbable or occur slowly within a human life span become inevitable and important in the context of geologic time. For example, catastrophic events that are improbable within the lifetime of an individual, such as meteorite impacts and ash flow eruptions eventually occur. Also, slow processes, like continental movement and mountain building that are barely perceptible on a human time scale, become significant. The measurement of absolute time by radiometric age dating, and the principles by which relative age determinations are made, are described.

We then describe the processes by which organisms become fossils. Most Earth science and introductory geology texts omit discussions of fossils and evolution; the topic is reserved for courses in historical geology. We feel that this omission is a mistake. Paleontology is a fascinating subject, it is an important part of geology, and it is an important tool in dating and correlating rocks and studying sedimentary environments. In this chapter, therefore, we discuss fossils, the history of life, and mass extinctions as integral parts of Earth science and the Earth's history.

Recently it has been suggested that meteorite impacts are not only the cause of mass extinctions, but may initiate continental rifting, the movement of tectonic plates, and are also important in the development of other large-scale geologic features such as basalt plateaus. This topic is discussed in this chapter in the context of extinctions, and in other chapters in the contexts of plate tectonics and flood basalts.

An interesting addendum to the meteorite impact story is that cosmic collisions may have helped create life as well as destroy it. Meixun Zhao and Jeffrey Bada of Scripps Institution of Oceanography reported finding unusually high concentrations of two amino acids, amino-isobutyric acid and isovaline in the Cretaceous-Tertiary boundary sediment. These amino acids are rare on Earth but have been detected in meteorites. If complex organic compounds were transported to Earth by collision with extraterrestrial objects, it is possible that these compounds were instrumental in the evolution of life.

A brief explanation of the processes and tools of geologic correlation leads into descriptions of the geologic time scale and the units of geologic time. The chapter ends with descriptions of the great length of geologic time, and a succinct review of the Earth and life through time.

Two special topics boxes review carbon-14 age dating, and modern extinctions.

Answers to Discussion Questions

1. Many such analogies can be constructed. Some can be based on familiar units of measure, such as a mile, a kilometer, a day, a year, etc. Others can be local examples, such as the distance to the next town, from a fraternity house to the nearest sorority, and so forth.

2. Assuming that the radiometric date on the biotite is unaffected by weathering, (a) would give the cooling age of the granite; (b) would give the time that the schist cooled through the argon retention temperature of the biotite (assuming that metamorphic temperature initially rose above that temperature); (c) would give the age of the igneous or metamorphic event during which the biotite formed, not the age of deposition of the sand. The biotite must be detrital, since it rarely or never forms in sedimentary environments. In all three cases it is necessary to compensate for initial argon uptake during formation of the biotite. In general, this question can be used to get into as detailed a discussion of the complexities of radiometric dating as the professor wishes.

3. Mineralization produces a precise replica of shells and bones, but does not normally preserve much evidence regarding the nature of soft tissue. Some fossils formed by preservation, such as mammoths frozen in permafrost contain intact organs. DNA has been extracted from cells of frozen mammoths. This type of information is not available in mineralized fossils.

4. From a single set of tracks, one could infer size, mass, whether the animal moved on two or four feet, whether it dragged its tail or held it up, whether it was a runner or sluggish, and other characteristics of its mobility. If you found many sets of identical tracks, you might infer that the species traveled in herds. If you found nests containing numbers of young animals, such as those that have been found in northern Montana, you might conclude that the animals nurtured their young.

5. An angular unconformity records deposition of sedimentary layers followed by tectonic tilting of those layers, erosion, and then deposition of a second set of layered sediment on the erosional surface.

A disconformity records deposition of sedimentary layers followed by either (a) a long time during which no further deposition occurred; or (b) uplift without tilting, and erosion of some of the layers. Later, more sedimentary layers are deposited on top of the older ones. Because no tilting occurred, the upper layers are parallel to the lower ones, but a long time interval exists between the two sets of sedimentary layers.

A nonconformity occurs when igneous or metamorphic rocks form below the Earth's surface and are then uplifted and eroded. Finally sedimentary layers are deposited on the erosion surface.

6. Lithologic correlation probably would be useless because of the great distance between the two localities. Time equivalency of the two sections might be established by correlation based on fossil assemblages, particularly index fossils; content of the same key beds, such as bentonite; or radiometric

dating of authigenic minerals such as glauconite, or of biotite or sanidine in bentonite.

7. The Phanerozoic eon is divided into the Paleozoic, the Mesozoic, and the Cenozoic eras. As the Greek roots imply, the three eras are distinguished by the life forms found in the corresponding rocks. In the Paleozoic (old life) era, life progressed from marine invertebrates to fishes, amphibians, and then reptiles. Land plants also appeared and evolved during the Paleozoic era. Dinosaurs appeared early in the Mesozoic era, and became dominant later in Mesozoic time. Flowering plants also appeared in the Mesozoic era. Mammals first appeared near the end of the Mesozoic era, and became dominant during the Cenozoic era. Grasses appeared in Cenozoic time, and became important food for grazing mammals. Sedimentary rocks that formed during each of these eras predominantly contain fossils of the animals and plants that dominated the era.

Periods are subdivisions of eras, and consequently are shorter units of time. For example, the Cambrian period is the first subdivision of the Paleozoic era. Sedimentary rocks of any period, like those of an eon, contain fossils of animals and plants that existed during that time. Thus, sedimentary rocks of each period are characterized by a particular assemblage of fossils. The periods of the Phanerozoic era differ in length because the periods were originally defined on the basis of fossil assemblages, and each group of animals and plants characteristic of the rocks of a single period existed for different lengths of time.

Although the boundaries separating eras and periods were originally based on fossils, and on strata that occur in particular parts of the world, most geologists now use time boundaries expressed in years, based on radiometric dating, in discussing the divisions of geologic time.

8. According to the results of several experiments, life must have evolved in an environment in which the temperature was between 0° and 100° C, and in an aqueous medium containing abundant compounds of carbon, oxygen, hydrogen, and other building blocks of organic molecules. In addition, an energy source, such as lightening, must have been available to initiate reactions among the compounds. Thus, many Earth scientists envision an Archean marine environment, at least in some places, characterized by shallow, warm, water rich in dissolved ions and, perhaps, suspended molecules; and with a stormy atmosphere.

Selected Bibliography

Keith Allen and Derek Briggs: Evolution and the Fossil Record. Washington, DC, Smithsonian Press, 1989. 265 pp.

Stephen Jay Gould: Time's Arrow Times' Cycle, Myth and Metaphor in the Discovery of Geological Time. Cambridge, Massachusetts, Harvard University Press, 1987. 222 pp.

G. Faure: Principles of Isotope Geology. New York, John Wiley & Sons, Inc., 1977.

A. Palmer: The Decade of North American Geology 1983 Geologic Time Scale. Geology 11:503-504.

Dale A. Russell: The Dinosaurs of North America: An Odyssey in Time. Toronto, University of Toronto Press, 1989. 239 pp.

Steven M. Stanley: Earth and Life Through Time 2nd ed. Baltimore, Johns Hopkins University Press, 1989. 689 pp.

An account of the Burgess shale fossils is given in:
Stephen Jay Gould: Wonderful Life, The Burgess Shale and Nature of History. New York, W. W. Norton, 1989. 347 pp.

S. Conway Morris: Burgess Shale Faunas and the Cambrian Expression. Science, 246:339, 1989.

An excellent review of fossils is available in this readable book:
George Gaylord Simpson: Fossils and the History of Life. New York, Scientific American Press, 1983. 239 pp.

Four entertaining articles on evolution are:
Stephen Jay Gould: Not Necessarily a Wing. Natural History, 10:12, 1985.

Stephen Jay Gould: Life's Little Joke. Natural History, 16, April 1987.

Roger Lewin: A Lopsided Look at Evolution. Science, 241:291, 1988.

Mark A. S. McMenamin: The Emergence of Animals. Scientific American, 94, April 1987.

The meteorite impact theory of mass extinctions has been debated in the literature. A sampling of some of the articles written at a non-technical level include:
Thomas J. Crowley and Gerald R. North: Abrupt Climate Change and Extinction Events in Earth History. Science, 240:996, 1988.

Anthony Hallam: End-Cretaceous Mass Extinction Event: Argument for Terrestrial Causation. Science, 238:1237, 1987.

John Horgan: The Impact Giveth. Scientific American, September 1989.

Les Kaufman and Kenneth Mallory, (eds.): The Last Extinction. Cambridge, Massachusetts, The MIT Press, 1986. 208 pp.

Richard A. Kerr: Huge Impact is Favored K-T Boundary Killer. Science, 242:865, 1988.

Digby J. McLaren and Wayne D. Goodfellow: Geological and Biological Consequences of Giant Impacts. Annual Review of Earth and Planetary Sciences 18 (123), 1990.

Chapter 4

Geologic Time, Fossils, Evolution, and Extinction

Multiple Choice:

1. The principle that states that geological processes and scientific laws operating today also operated in the past is called (a) uniformitarianism; (b) the principle of superposition; (c) principle of horizontality; (d) the principle of cross-cutting relationships.

2. Absolute time tells us
(a) the order in which events occurred; (b) the amount of time separating events; (c) the order in which events occurred without regard to amount of time separating them;(d) both a and b.

3. Measurement of absolute time depends on
(a) measuring the constant rate of a process; (b) measuring the cumulative effects of a process; (c) discovering fossils; (d) measuring unconformities (e) a and b.

4. Isotopes are atoms of the same element with ______ numbers of neutrons.
(a) different; (b) the same; (c) opposite

5. The half-life of K-40
(a) refers to the fact that half the world's reserves of K-40 are already depleted; (b) refers to the amount of time needed for one-half of a quantity of K-40 to decay; (c) varies with chemical conditions; (d) varies with time; (e) varies with depth in the crust.

6. Carbon-14 gives accurate ages of materials younger than _____ years.
(a) 4 billion years; (b) 500 million years; (c) 70,000 years; (d) 10,000 years.

7. If layers of sedimentary rocks are deposited without interruption, they are called
(a) discontinuous; (b) an unconformity; (c) a nonconformity; (d) conformable.

8. An interruption of deposition of sediment, usually of long duration is called
(a) a fossil; (b) an unconformity; (c) the principle of horizontality; (d) the principle of superposition.

9. A basalt dike cutting through sedimentary rock illustrates (a) a fossil; (b) the principle of cross-cutting relationships; (c) the

principle of horizontality; (d) the principle of superposition

10. When secondary minerals fill the impression of a fossil, they form
(a) a cast; (b) an internal mold; (c) an external mold; (d) a trace fossil; (e) a permineralized fossil.

11. Dinosaur tracks are examples of
(a) casts; (b) internal molds; (c) trace fossils; (d) replacement fossils; (e) recrystallized fossils.

12. The method of interpreting relative time by studying the fossils found in sedimentary rock is called
(a) the principle of faunal succession; (b) the principle of cross-cutting relationships; (c) the principle of horizontality; (d) the principle of superposition.

13. The matching of rocks of similar ages from different localities to assemble a history of the Earth is called
(a) the principle of horizontality; (b) a fossil; (c) correlation (d) the principle of superposition.

14. The largest units of geologic time are
(a) eras; (b) epochs; (c) eons; (d) periods; (e) columns

15. Which eon means "earlier life"?
(a) Phanerozoic; (b) Proterozoic; (c) Paleozoic; (d) Mesozoic; (e) Cenozoic.

16. The Phanerozoic eon is so finely divided because
(a) it was the longest eon; (b) it makes up over 90 percent of Earth history; (c) fossils are abundant in rocks of the eon; (d) Phanerozoic rocks are younger and more commonly well preserved; (e) both c and d.

17. In early Paleozoic time, life was almost completely confined to
(a) the land; (b) the oceans; (c) chalk beds; (d) swampy areas

18. All species of dinosaurs suddenly became extinct
(a) 65 million years ago; (b) 4.6 billion years ago; (c) 5 million years ago; (d) 10 million years ago.

19. The latest of the four eras which began 66 million years ago is the____ era.
(a) Phanerozoic; (b) Proterozoic; (c) Paleozoic; (d) Mesozoic; (e) Cenozoic

20. Primitive humans evolved during the Cenozoic era, between
(a) 500,000 and 1 million years ago; (b) 4.6 billion years ago; (c) 500,00 million years ago; (d) 2000 years ago.

21. During the _____ era there was a sudden explosion in the number of species of multi-cellular organisms with hard shells which were preserved as fossils.
(a) Phanerozoic; (b) Proterozoic; (c) Paleozoic; (d) Mesozoic; (e) Cenozoic

22. During the _____ era fish and shelled organisms were abundant but there were few land animals.
(a) Phanerozoic; (b) Proterozoic; (c) Paleozoic; (d) Mesozoic; (e) Cenozoic

True or False:

1. James Hutton's conclusions about geological change were based on a principle now known as uniformitarianism.

2. Uniformitarianism states that geological processes and scientific laws operating today did not operate in the past.

3. Measurement of relative time is based on the principal that in order for an event to affect a rock, the rock must exist first.

4. Potassium-40 decomposes naturally to form two other isotopes, argon-40 and calcium-40.

5. An isotope created by radioactive decay of a parent, such as argon-40 or calcium-40, is called a parent isotope.

6. The quantity of daughter isotopes in a sample increases with time.

7. Younger layers of sediment always accumulate below older layers.

8. In an angular unconformity, older sedimentary beds were tilted and eroded before younger layers were deposited.

9. It is more common for fossils an organism to be formed by mineralization than by preservation.

10. The subdivisions of geologic time are based largely on fossils found in rocks that formed during each interval.

11. The Earth's history is divided into 6 eons.
12. The word Phanerozoic derives from ancient Greek roots meaning evident life.

13. Hadean time has abundant fossils.

14. At the end of the Proterozoic eon, and the start of the Phanerozoic eon, large, multicellular plants and animals evolved.

15. The end of Paleozoic era was marked by a mass extinction in which half of all families of organisms died.

16. One theory for the death of the dinosaurs is that a giant meteorite struck the Earth.

Completion:

1. _____ _____ expresses the order in which rocks and features form.

2. _______ _______ is measured in years.

3. _____ are atoms with the same number of protons but different numbers of neutrons.

4. Measurement of time using natural radioactivity is _____ ____ _____.

5. The principle of _____ _____ is based on the observation that sediment usually accumulates in horizontal beds.

6. The principle of _____ states that sedimentary rocks usually become younger from bottom to top.

7. A basalt dike cutting sedimentary rocks is an example of the principle of _____ _____ _______.

8. ______ ______ are not remains of an organism, but instead are tracks, burrows, or other marks made by the organism.

9. Use of fossils to tell the relative age of a rock employs the principle of _____ _____.

10. Matching rocks of similar ages from different localities is _____.

11. A _______ ______ is a diagram of rocks from all over the world which represents a continuous record of most of the Earth's history.

12. The smallest time units of the geologic time scale are called ______.

13. The earliest time in Earth's history is the ______ eon.

14. Large multicellular plants and animals evolved abruptly at the start of the _____ eon.

15. Primitive men and women evolved during the _____ eon.

Answers for Chapter 4

Multiple Choice 1. a; 2. c; 3. e; 4. a; 5. b; 6. c; 7. d; 8. b; 9. b; 10. b; 11. c; 12. a; 13. c; 14. c; 15. b; 16. e; 17. b; 18. a; 19. e; 20. a; 21. a; 22. c

True or False 1. T; 2. F; 3. T; 4. T; 5. F; 6. T; 7. F; 8. T; 9. T; 10. T; 11. F; 12. T; 13. F; 14. T; 15. T; 16. T

Completion 1. relative time; 2. absolute time; 3. isotopes; 4. radioactive dating; 5. original horizontality; 6. superposition; 7. cross-cutting relationships; 8. trace fossils; 9. faunal succession; 10. correlation; 11. geologic colummn; 12. epochs; 13. Hadean; 14. Phanerozoic; 15. Cenozic

CHAPTER 5

Plate Tectonics

Discussion

This chapter is the first of five chapters, 5 through 9, that describe the Earth's internal processes, concentrating on the plate tectonics theory and the geologic consequences of movements of tectonic plates. Chapter 5 describes the main aspects of the theory. Chapter 6 discusses the causes and nature of earthquakes in terms of the model. Chapter 7 describes the origins of magma, volcanoes, and plutons in the context of plate tectonics. Chapter 8 introduces geologic structures and then uses the plate tectonics model and structural geology to discuss the origins of mountains and mountain ranges. Chapter 9 applies the plate tectonics model, as well as concepts learned in earlier chapters, to describe the geological evolution of North America.

In Chapter 5 we discuss the basic elements of the theory of plate tectonics. Plate tectonics is such an important concept that it revolutionized the science of geology. We use an historical approach to illustrate the evolution of a major idea in science. We feel that it is not only important for students to learn the theory, but also to understand the process of scientific thought and the manner in which a theory is constructed from observations and data.

The chapter begins with a description of Wegener's evidence for continental drift, the development of his model, and a discussion of why his work was rejected by most of the scientific community. We then discuss the ocean floor magnetic evidence that resulted in the model of sea-floor spreading and the rapid expansion of that model into the theory of plate tectonics. The most important elements of the plate tectonics theory, and the anatomy of tectonic plates are described. The chapter concludes with a discussion of why tectonic plates move.

Chapter 5 includes a special topics box describing an hypothesis that relates meteorite impacts to the initiation of lithospheric rifting and the origin of basalt plateaus.

Answers to Discussion Questions

1. Wegener's theory and the modern plate tectonics theory have nearly identical descriptions of a supercontinent, Pangaea, at the end of Triassic time. In both theories, Pangaea began to split up in late Triassic time, and the continents continued to separate and travel in much the same paths to their present locations.

Wegener was primarily interested in the evidence that Pangaea had once existed and then split up into the present continents. He suggested mechanisms for continental drift only as an afterthought to his theory. His explanation was that continents either plough through, or slide over, oceanic crust. Both mechanisms were quickly proved impossible because of rock properties. In contrast, the mechanism of movement of lithospheric plates by gliding over the plastic asthenosphere is an essential, and mechanically plausible, part of the plate tectonics theory.

Wegener's supporting data consisted of coastline congruencies, fossil and paleoclimatic evidence, and continuity of geologic features in reconstructions of Pangaea. The plate tectonics theory is based on paleomagnetic, seismic, heat flow, and high-technology survey data, as well as Wegener's evidence and additional evidence of the kinds found by Wegener.

2. Section 5-10 discusses mechanisms for plate movement, and concludes that a combination of several models seems plausible, although most geologists do not agree on the subject. This is a good time to reinforce the idea that the development of the plate tectonics theory is an on-going and active field of research.

3. Strong earthquakes may occur within lithospheric plates as a result of transfer of forces from an active plate margin through the lithosphere to the plate interior. They may also result from differential vertical movement of a plate interior as it passes over a mantle plume, or a "bump" in the asthenosphere.

4. Impact of a large meteorite on the Earth's surface would have several major effects described in the "Focus On" box. Climates would change as dust blocks solar radiation. The climate changes might cause mass extinctions. A large crater would form, and perhaps fill with basalt to form a basalt plateau. The impact might initiate lithospheric rifting, and perhaps breakup of a continent.

Some possible effects are not described in the box. A large basalt plateau on a continent is a region of anomalously high density in contrast to normal continental crust. Eventually such a region might sink isostatically to form an intracratonic sedimentary basin. This may be how such basins form. Major earthquakes would be triggered. If a large meteorite fell into the ocean, energy from its impact might vaporize vast quantities of seawater, raising the relative humidity of the entire global atmosphere to 100 percent. This is enough moisture to generate 7 to 13 meters of rain globally in a single storm. A deluge of Noachian proportion would follow, causing erosion, transport, and deposition of sediment that should be recognizable in the geologic record. The possibilities are endless. An imaginative class can have a lot of fun with this one!

5. A large meteorite impact on the moon would generate major moonquakes, and would certainly throw up a lot of dust. However, the moon is probably too cool, and its crust and lithosphere too thick, for igneous activity to be triggered, or for lithospheric plate motion to be initiated by an impact.

Selected Bibliography

An excellent little book which gives Wegener's own version of his story is:
A. Wegener: The Origins of Continents and Oceans. Translated from the 4th revised German edition, 1929, by J. Biram. London, Methuen, 1966.

General references on Tectonics are given in:
A. Hallam: A Revolution in the Earth Sciences: From Continental Drift to Plate Tectonics. London, Oxford University Press, 1973.

S. Uyeda: The New View of the Earth: Moving Continents and Moving Oceans. New York, W. H. Freeman, 1978. 217 pp.

P. Hoffman: Speculations on Laurentia's First Gigayear (2.0 to 1.0 Ga). Geology, 17 135-138.

S. Carey: Theories of the Earth and Universe: a History of Dogma in the Earth Sciences. Stanford, CA, Stanford University Press, 1988.

A. Cox and R. Hart: Plate Tectonics. London, Blackwell Scientific Publishers, 1986. 400 pp.

P. Davies and S. Runcorn, eds.: Mechanisms of Continental Drift and Plate Tectonics. New York, Academic Press, 1980.

Chapter 5

Plate Tectonics

Multiple Choice:

1. Alfred Wegener postulated the theory of
(a) uniformitarianism; (b) continental drift; (c) catastrophism; (d) plate tectonics; (e) seismic discontinuity.

2. Alfred Wegener proposed his theory from what type or types of evidence
(a) fit of continental coastlines; (b) fossil evidence; (c) paleoclimatology; (d) a and b; (e) a, b, and c.

3. Wegener suggested that the Atlantic Ocean began to open in _____ time.
(a) Cretaceous; (b) Triassic; (c) Precambrian; (d) Ordovician; (e) Tertiary

4. Physicists rejected Wegener's theory because
(a) the fit of Africa and South America was wrong; (b) the timing of opening of the Atlantic was off by 200 million years; (c) Mesosaurus was a swimming animal; (d) his mechanism for making continents drift was untenable; (e) all of the above

5. Iron-bearing minerals in igneous rock can be used to
(a) record the magnetic orientation of the Earth's magnetic field at the time the rock cooled; (b) give the age of the rock; (c) prove that the Earth's magnetic field has not reversed over time; (e) none of the above.

6. Sea-floor basalt
(a) has a magnetic orientation pointing north, east of the Mid-Atlantic ridge and pointing south, west of the Mid-Atlantic ridge; (b) has a magnetic orientation pointing south, east of the Mid-Atlantic ridge and pointing north, west of the Mid-Atlantic ridge; (c) has alternating bands of north and south magnetic orientation; (d) has alternating bands of south-pointing magnetic orientation; (e) has alternating bands of north-pointing magnetic orientation.

7. The chemical composition of the mantle is
(a) different from place to place; (b) nearly constant throughout; (c) mainly basalt; (d) mainly granite.

8. The outer 100 kilometers of the Earth including both the crust and upper mantle is called the
(a) isosphere; (b) lithosphere; (c) asthenosphere; (d) Mid-Atlantic ridge.

9. The ocean floor is lower than the continents because
(a) the denser oceanic lithosphere settles down farther into the asthenosphere than the less dense granite of continental lithosphere does; (b) the oceanic lithosphere is thicker than continental lithosphere; (c) they are weighted down by seawater; (d) oceanic lithosphere is thinner than continental lithosphere.

10. The center of the Earth's core is
(a) cool and brittle; (b) composed of helium; (c) liquid because the pressure is very high; (d) very hot, like the surface of the Sun.

11. The lithosphere is
(a) hot and plastic; (b) mostly liquid magma; (c) composed of sedimentary rocks; (d) cool and brittle.

12. The asthenosphere is
(a) hot and plastic; (b) mostly liquid magma; (c) composed of sedimentary rocks; (d) cool and brittle.

13. In the northern hemisphere, beaches from the Ice Ages that are now high above sea level demostrate the principle of
(a) normal polarity; (b) reversed polarity; (c) isostatic adjustment; (d) sea-floor spreading; (e) continental rifting.

14. The theory of plate tectonics presents a model of Earth in which
(a) brittle lithospheric plates are solidly attached to the asthenosphere and do not move; (b) the lithosphere is broken into 7 large segments and several smaller ones; (c) lithospheric plates glide slowly over the asthenosphere; (d) b and c.

15. When lithospheric plates collide they
(a) generate force great enough to build mountains and cause volcanic eruptions and volcanoes; (b) bump into each softly and change direction; (c) create no visible effects; (d) buckle and thus make the Earth's diameter increase; (e) make a continuous rumble, like the roar of surf.

16. A divergent boundary in oceanic crust forms
(a) a subduction zone; (b) the mid-oceanic ridge; (c) an oceanic trench; (d) a Benioff zone.

17. The mid-oceanic ridge is elevated above surrounding sea floor because
(a) it is made of the newest, hottest, and lowest-density lithosphere; (b) it is made of the oldest, coolest, and most dense lithosphere; (c) it is full of gas bubbles; (d) it is very old.

18. When two plates move horizontally toward each other and collide, they form a
(a) divergent boundary; (b) a rift zone; (c) a convergent boundary; (d) a transform boundary; (e) a mid-oceanic ridge.

19. When a continental plate collides with an oceanic plate,
(a) the oceanic plate floats above the continental plate (b) the continental plate sinks into the mantle; (c) the oceanic plate sinks beneath the continental plate and dives into the mantle; (d) a divergent boundary forms.

20. New crust formed at spreading centers
(a) is consumed at mid-ocean ridges; (b) proves that the earth is expanding; (c) is consumed at hot-spots on other plates; (d) is lost on transform faults; (e) is balanced by subduction elsewhere.

21. Subduction occurs
(a) at mid-ocean ridges; (b) along transform faults; (c) at hot-spots; (d) at convergent plate boundaries; (e) along continental rifts.

22. Huge quantities of magma form in subduction zones because
(a) addition of water from sea floor mud and basalt causes the rock to melt; (b) the asthenosphere cannot melt at a subduction zone; (c) large concentrations of uranium at subduction zones heat the rock to melting by radioactive decay; (b) a and c; (e) none of the above.

23. Many of the world's great mountain chains, including the Andes and parts of the mountains of western North America, form because
(a) the great volume of magma rising through the Earth's crust thickens the crust, causing mountains to rise; (b) volcanic eruptions pour huge amounts of lava onto the surface, constructing chains of volcanoes; (c) the Earth's crust crumples and buckles into mountain ranges where two lithospheric plates crash together; (d) a, b, and c; (e) none of the above.

24. The San Andreas Fault is an example of
(a) a convergent boundary; (b) a divergent boundary; (c) a transform boundary; (d) a Benioff zone; (e) a rift valley.

25. Tectonic plates move away from spreading centers at
(a) 1 to 18 centimeters per year; (b) 1 to 18 kilometers per year; (c) less than 1 centimeter per year; (d) fast enough to be watched by humans.

26. A portion of a plate with continental crust is ______ than a portion of a plate with oceanic crust.

(a) thicker and denser; (b) thicker and less dense; (c) thinner and denser; (d) thinner and less dense

True or False:

1. A fault is any fracture in rock along which movement has occurred.

2. Alfred Wegener developed the theory known as continental drift.

3. The Earth's magnetic field has never reversed polarity throughout geologic history.

4. The world's largest and longest mountain chain is the mid-oceanic ridge.

5. Sea-floor basalt cannot be used to record the orientation of the Earth's magnetic field because it is rich in iron.

6. As the new sea floor cools, it acquires the orientation of the Earth's magnetic field.

7. Oceanic crust ranges from 100 to 200 kilometers in thickness and is composed mostly of granite.

8. Continental crust is much thicker than oceanic crust.

9. Most earthquakes originate in the brittle rock of the lithosphere.

10. The asthenosphere is a rigid, brittle rock layer.

11. The lithosphere floats on the asthenosphere.

12. As Earth melted, most of the heavy elements gravitated away from the core which is made up of light elements.

13. The mid-oceanic ridge is elevated above surrounding sea floor because it is made of the newest, hottest, and lowest-density lithosphere.

14. The deepest point on Earth is in the Mariana trench, in the southwestern Pacific Ocean north of New Guinea, where the sea floor is as much as 10.9 kilometers below sea level.

15. If two colliding plates are both covered with continental crust, subduction will occur.

16. A plate is a segment of the lithosphere; thus, it includes the uppermost mantle and all of the overlying crust.

17. A mantle plume is a vertical column of plastic rock rising through the mantle.

Completion:

1. Wegener proposed that all the world's continents where fitted together in one supercontinent he called __________.

2. Wegener mapped the location of ________ remains, as additional evidence for his theory of continental drift.

3. The orientation of the Earth's field at present is referred to as normal, and that during a time of opposite polarity is called ______.

4. The ______ ______ is a mountain range in the middle the Atlantic Ocean basin, halfway between Europe and Africa to the east and North America and South America to the west.

5. The ______ is almost 2900 kilometers thick and makes up about 80 percent of the Earth's total volume.

6. The core consists mainly of iron and ______.

7. The theory of ______ explains why the ocean floor is lower than the continents.

8. A continent can be pulled apart at a divergent boundary in a process called ______ ______.

9. Elongate depressions called ______ ______ develop along continental rifts because continental crust becomes stretched, fractured, and thereby thinned as it is pulled apart.

10. As a lithospheric plate sinks into the mantle during subduction, it slips and jerks causing numerous earthquakes; this zone of quakes is called the zone.

11. An ______ ______ is a long, narrow trough in the sea floor formed where a subducting plate turns downward to sink into the mantle.

12. A chain of volcanic islands rising from the sea floor is called a(an) ______ ______.

13. ______ ______ theory provides a mechanism for the movement

of continents in Wegener's continental drift theory.

14. A(an) ______ ______ ______ forms where two plates slide horizontally past each other.

15. a portion of a plate with continental crust composing its uppermost layer is ______ than one bearing oceanic crust.

16. Earthquakes and faulting are common at all ______ ______.

17. Hot rock from deep in the mantle rises to the base of the lithosphere while cooler upper mantle rock sinks. This flow of solid rock is called ______ ______.

Answers for Chapter 5

Multiple Choice: 1. b; 2. e; 3. b; 4. d; 5. a; 6. c; 7. b; 8. b; 9. c; 10. d; 11. d; 12. a; 13. c; 14. d; 15. b; 16. a; 17. c; 18. c; 19. e; 20. d; 21. a; 22. d; 23. c; 24. a; 25. a; 26. b

True or False: 1. T; 2. T; 3. F; 4. T; 5. F; 6. T; 7. F; 8. T; 9. T; 10. F; 11. T; 12. F; 13. T; 14. T; 15. F; 16. T; 17. T

Completion: 1. Pangaea; 2. fossil; 3. reversed; 4. Mid-Atlantic ridge; 5. mantle; 6. nickel; 7. isostacy; 8. continental rifting; 9. rift valleys; 10. Benioff; 11. oceanic trench; 12. island arc; 13. plate tectonics; 14. transform plate boundary; 15. thicker or less dense; 16. plate boundaries; 17. mantle convection

CHAPTER 6

Earthquakes and the Earth's Structure

Discussion

Earthquakes fascinate Earth science students, at least partly because they are manifestations of the tectonically active Earth that occur frequently and impressively enough to be in the headlines at least once during each quarter- or semester-long course. In this chapter we use the principles of plate tectonics taught in Chapter 5 to explain how, why, and where earthquakes occur, and use several examples of recent newsworthy quakes, including those that shook both northern and southern California in early 1992.

In the first several sections of Chapter 6, we describe earthquake and earthquake waves, and explain why and how most quakes occur along tectonic plate boundaries. Sections 6.3 and 6.4 explain how seismographs work, and how the epicenter of an earthquake is located. Sections 6.5 and 6.6 describe the effects of earthquakes on humans, and discuss the art and science of quake prediction. In Section 6.7 we describe how the study of seismic waves reveals the Earth's layered structure.

We describe earthquake history, potential, and tectonic setting for most parts of North America in special topics boxes on "Earthquake Danger in the Central and Eastern United States," and "Two Hundred Years of Earthquake Activity in the San Francisco Bay Area," and in Section 6.2 with a discussion of subduction zone earthquake history and potential in the Pacific Northwest.

In section 6.5 we explain that earthquake damage is partly related to the quality of construction in an affected area. It is misleading to make exactly parallel comparisons when measuring quake damage. Yet, when comparing the 1988 Armenian quake (Richter magnitude 6.9, 24,000 people killed) and the Iranian quake (Richter magnitude 7.7, 40,000 people killed), with the San Francisco quake (Richter magnitude 6.9, 65 people killed), it is certain that quality of construction was a crucial factor in preventing greater loss of life in the San Francisco quake.

Construction codes in earthquake zones are a good example of "risk analysis". In making a risk analysis, one must evaluate both the probability of an event and the consequences of the event. Therefore, an area that might be considered "safe" for home construction might not be safe for a nuclear power plant. The difference arises because the consequences of failure of the nuclear power plant are so much greater they are for private homes.

The plastic behavior of rocks confuses some students because it contradicts their intuitive sense of solid rock. A professor might use the analogy of sagging window glass in very old houses to show that a solid can be both fluid and brittle. Similarly, if a ball of tar is cooled in a freezer and struck with a hammer, it fractures. However, if heated, the tar ball can be deformed plastically by hand. On a hot summer day the kickstand of a bicycle sinks into a tarry road. A simple classroom demonstration using road tar can also illustrate the point. If a ball of tar is placed in a wide-necked funnel and left, the tar slowly oozes downward. In a few months a drop will form and if the consistency is right and the temperature high enough, the drop will fall before the final week of the course.

Answers to Discussion Questions

1. Earthquakes occur when rock fractures. Plastic deformation produces no sudden movement and therefore does not initiate seismic waves.

2. 3600 kilometers.

3. Northwest Wyoming, in Yellowstone National Park.

4. The question asks students to integrate geology with public policy and legal responsibility. If the fluid injection works and a large earthquake could be averted, the city would benefit. However, the technology of earthquake control is not precise enough to guarantee success. If injected fluids initiated an earthquake that led to property damage and loss of life, the political and legal liability would be overwhelming.

5. The intervals between quakes were 24, 20, 21, 12, and 32 years, with an average interval of 22 years. The laws of probability do not allow us to make a precise prediction, but another Parkfield earthquake might be expected in 1989 $\pm$ 10 years. Statistical correlations of this type provide a general reference for evaluating earthquake danger, but they are not precise enough for short-term prediction.

6. It would be enormously expensive to close down a large city such as San Francisco, and evacuate its residents.

(upper left hand box) If a predicted earthquake occurs and the city is evacuated, lives will be saved and the geologist who predicted it would be applauded.

(upper right hand box) However, if an earthquake is predicted, the city is evacuated, and the quake does not occur, those responsible might be criticized and even sued.

(lower left hand box). If a geologist were to predict a quake, the prediction were ignored, and the quake occurred, everyone would be asked why the prediction was ignored.

(lower right hand box) If an earthquake was not predicted and the city was not evacuated, the lack of prediction would not be particularly noteworthy.

An interesting parallel exists between this problem and industrial accidents. Just before the explosion of the chemical plant in Bhopal, India, the control room operator noted an alarming build up of pressure in the tank where the methyl isocyanate was stored. At this point quick action could have averted disaster, but the operator would have had to destroy the contents of the tank. The operator feared that perhaps the pressure gauge was erratic, and that his superiors would censure him for acting if no emergency existed. So the man did a human thing, he ran out to ask his supervisor. By the time the two men had returned to the control room, it was too late.

In making decisions, you can't always be right, but no one wants to cry wolf unless he or she is certain a real wolf threatens, and often by then it is too late. The point here is that the penalties for evacuating the city incorrectly are much greater than the rewards for making a correct and life-saving decision.

Selected Bibliography

Reference material on earthquakes:
Bruce A. Bolt: Earthquakes. New York, W. H. Freeman, 1988. 282 pp.

Cliff Frohlich: The Nature of Deep-Focus Earthquakes. Annual Review of Earth and Planetary Sciences 19 (227), 1989.

Cliff Frohlich: Deep Earthquakes. Scientific American, (48), January 1989.

Thomas H. Heaton and Stephen H. Hartzell: Earthquake Ground Motions. Annual Review of Earth and Planetary Sciences 16 (121), 1988.

Kiyoo Wadati: Born in a Country of Earthquakes. Annual Review of Earth and Planetary Sciences 17 (1), 1989.

Earthquake hazards in California:
W. H. Bakun and A. G. Lindh: The Parkfield, California, Earthquake Prediction Experiment. Science, 229:619, 1985.

Richard A. Kerr: Stalking the Next Parkfield Earthquake. Science, 223:36, 1984.

Richard A. Kerr: New Fault Picture Points Toward Bay Area Earthquakes. Science, 244:286, 1989.

Richard A. Kerr: Reading the Future in Loma Prieta. Science, 246:436, 1989.

Sandra S. Schultz and Robert E. Wallace: The San Andreas Fault. United States Geological Survey publication. 17 pp.

Robert L. Wesson and Robert E Wallace: Predicting the Next Great Earthquake in California. Scientific American, 252:35, 1985.

United States Geological Survey: Active Faults of California. USGS INF-74-3 (R.1)

Earthquake hazards in the United States outside of California:
Klaus H. Jacob and Carl J. Turkstra (eds.): Earthquake Hazards and the Design of Constructed Facilities in the Eastern United States. New York, New York Academy of Sciences, 1989. 455 pp.

Arch C. Johnston: A Major Earthquake Zone on the Mississippi. Scientific American, (60), April 1982.

The interior of the Earth and the Earth's magnetism are discussed in:
Jeremy Bloxham and David Gubbins: The Evolution of the Earth's Magnetic Field. Scientific American, (68), December 1989.

Vincent Coutillot and Jean Louis Le Mouel: Time Variations of the Earth's Magnetic Field: From Daily to Secular. Annual Review of Earth and Planetary Sciences 16 (389), 1988.

Craig M. Jarchow and George A. Thompson: The Nature of the Mohorovicic Discontinuity. Annual Review of Earth and Planetary Sciences 19 (475), 1989.

Raymond Jeanloz: The Nature of the Earth's Core. Annual Review of Earth and Planetary Sciences 18 (357), 1990.

Seismic tomography:
Don L. Anderson and Adam M. Dziewonski: Seismic Tomography. Scientific American, (60), 1988.

R. K. O'Nions and B. Parsons, (eds.): Seismic Tomography and Mantle Circulation. London, Royal Society Press, 1989. 152 pp.

Chapter 6

Earthquakes and the Earth's Structure

Multiple Choice:

1. How many earthquakes occur each year?
(a) 1 million; (b) 10,000; (c) 500; (d) 2 dozen; (e) one or two.

2. As tectonic plates glide over the hot, plastic asthenosphere, they slip past one another along immense fractures called
(a) plate boundaries; (b) bedding planes; (c) dikes; (d) epicenters; (e) none of the above.

3. The edges of two plates may remain stationary while the interiors of the plates move because
(a) rock is brittle; (b) rock is elastic; (c) the plates are fused solidly; (d) rock is non elastic.

4. In most cases an earthquake will start at
(a) a mountain top; (b) the interior of a large continent; (c) an old fault; (d) bedding planes (e) dikes.

5. Western California is moving northwest at the rate of _____ relative to eastern California.
(a) 100 cm/yr: (b) 50 cm/yr; (c) 3.5 cm/yr; (d) .2 cm/yr; (e) 15 mph

6. The earthquakes caused when a subducting plate sinks into the mantle usually occur along
(a) the lower part of the sinking plate; (b) the middle part of the sinking plate; (c) the upper part of the sinking plate; (d) under the mantle.

7. Waves that travel through rock are called
(a) cosmic waves; (b) earth waves; (c) regular waves; (d) seismic waves; (e) radio waves.

8. A primary wave
(a) forms by alternate compression and expansion of rock; (b) travels much faster than the speed of sound; (c) passes through both solids and liquids; (d) is the first body wave to reach an observer; (e) all of the above; (f) none of the above.

9. The starting point of an earthquake where rocks move and body waves radiate in concentric spheres is the
(a) epicenter; (b) fault; (c) focus; (d) seismic; (e) elastic limit.

10. S-waves

(a) move only through solids; (b) are slower than p-waves; (c) travel at 3 to 4 km/second in the crust; (d) are shear waves; (e) all of the above; (f) none of the above.

11. To make a time-travel curve for an earthquake,
(a) only the time of the quake must be known; (b) only the location where the quake started must be known; (c) both the time and location of the quake must be known; (d) only one seismic station is required.

12. A quantitative scale based on seismograph measurements rather than human evaluation of damage is called
(a) the Mercalli scale; (b) the Richter scale; (c) the intensity scale; (d) the Lewis scale; (e) the decibel scale

13. The damage caused by an earthquake depends on
(a) magnitude; (b) proximity to population centers; (c) the quality of building construction in the affected area; (d) all of the above; (e) none of the above.

14. Secondary dangers of earthquakes are
(a) fires from ruptured gas lines; (b) landslides; (c) tsunamis; (d) all of the above; (e) none of the above.

15. Tsunamis are caused by
(a) earthquakes on the sea floor; (b) huge tides; (c) fires; (d) submarines.

16. An immobile region of a fault bounded by moving segments is called
(a) the epicenter; (b) the focus; (c) a seismic gap; (d) a seismograph; (e) a seismogram.

17. Scientists use seismic waves to
(a) locate transition zones in the Earth's interior; (b) determine whether regions in the Earth's interior are solid or liquid; (c) locate the source on an earthquake; (d) all of the above; (e) none of the above.

18. Bending of waves is
(a) rarefaction; (b) refraction; (c) reflection; (d) deflection; (e) inflection.

19. No s waves pass through the center of the Earth because
(a) the asthenosphere is liquid; (b) the outer core is liquid; (c) S waves travel only through liquids; (d) S waves refract at the crust mantle boundary.

20. The boundary where a sudden change in velocity of seismic waves occurs between the crust and mantle is called

(a) a seismic gap; (b) the Mohorovicic discontinuity; (c) a bedding plane; (d) the strike zone; (e) none of the above.

True or False:

1. An average of about 10,000 earthquake fatalities occur every year.

2. When two tectonic plates move past one another, rock near the plate boundary often stretches and stores elastic energy.

3. An earthquake occurs when accumulated energy in rocks is released as tectonic plates slide past one another.

4. The least likely place for an earthquake to occur is at a plate boundary.

5. Rocks on opposite sides of the San Andreas Fault move past each other at a speed of 5 kilometers a year.

6. Only deep earthquakes occur along the mid-oceanic ridge.

7. During an earthquake, surface waves radiate away from the epicenter along the surface of the Earth like the waves that form when you throw a rock into a calm lake.

8. P waves in the Earth travel more than ten times slower than the speed of sound in air.

9. Surface waves travel more slowly than either type of body wave.

10. Scientists use the arrival times of different seismic waves from three different reporting stations to locate the epicenter of an earthquake.

11. The Mercalli scale measures the magnitude of an earthquake from the amplitude of the largest wave recorded on a seismograph.

12. The largest earthquakes ever observed had magnitudes of 8.5 to 8.7, about 900 times greater than the energy released by the Hiroshima bomb.

13. The New Madrid earthquake of 1811 altered the course of the Mississippi River.

14. P and S waves travel everywhere and can be detected anywhere on the planet.

15. The change from solid rock to molten liquid at the core mantle boundary creates a shadow zone between 105° and 140° where neither S nor P waves arrive from an epicenter.

16. Cool, brittle rock of the lithosphere carries seismic waves more effectively than the hot, soft rock of the asthenosphere.

Completion:

1. An ______ is a sudden motion or trembling of the Earth.

2. When two tectonic plates move past one another, rock near the plate boundary stretches and stores ______ ______.

3. A type of continuous, snail-like movement that occurs along the San Andreas Fault is called ______ ______.

4. Intermittent jerks and slips that occur when a lithospheric plate sinks below an oceanic plate give rise to_______.

5. The waves that travel through rock the most quickly are called ______ _______.

6. During an earthquake, ______ ______ radiate away from the epicenter along the surface of the Earth like the waves that form when you throw a rock into a calm lake.

7. ______ waves can travel only through solids.

8. Earthquakes are detected and measured with a device called a ______.

9. Geologists use ______ ______ to measure the distance to an earthquake epicenter by recording arrival times of the different types of waves.

10. The ______ ______ evaluates earthquake strength on human experience and building damage.

11. ______ are small earthquakes that precede a large quake by an interval ranging from a few seconds to a few weeks.

12. A wave ______ and sometimes reflects as it passes from one transmitting medium into another.

13. The Earth's outer core is _____.

14. The _______ is composed mostly of iron and nickel.

15. The behavior of ______ ______ as they pass through the Earth have allowed scientists to study the Earth's interior.

Answers for Chapter 6

Multiple Choice: 1. a; 2. a; 3. b; 4. c; 5. c; 6. c; 7. d; 8. e; 9. c; 10. e; 11. c; 12. b; 13. d; 14. d; 15. a; 16. c; 17. d; 18. b; 19. b; 20. b

True or False: 1. T; 2. T; 3. T; 4. F; 5. F; 6. F; 7. T; 8. F; 9. T; 10. T; 11. F; 12. T; 13. T; 14. F; 15. T; 16. T

Completion: 1. earthquake; 2. elastic energy; 3. fault creep; 4. earthquakes; 5. P waves; 6. surface waves; 7. Secondary or S; 8. seismograph; 9. seismograph; 10. Mercalli scale; 11. foreshocks; 12. refracts; 13. liquid; 14. core; 15. seismic waves

CHAPTER 7

Plate Tectonics, Volcanoes, and Plutons

Discussion

Modern igneous petrologists tell us that magma can form in three ways: increasing the temperature of rock; decreasing pressure on hot, solid rock (pressure relief melting); and adding water to hot, solid rock. Most students intuitively conclude that magma must form as a result of heating of rock until it melts. Although this seems to be a natural explanation, increasing temperature is probably not a common cause of initial magma generation in the upper mantle. Instead, pressure relief melting at rift zones and mantle plumes, and addition of water to asthenosphere rock above a subducting plate, are thought to be the most important primary mechanisms of magma generation.

Large amounts of magma form in three environments, rift zones, mantle plume - hot spots, and subduction zones. Basaltic magma forms at rift zones by pressure relief melting of hot asthenosphere as it rises beneath separating lithospheric plates. Basaltic magma forms by the same pressure relief process above mantle plumes. In both of these cases, decreasing pressure is thought to be the primary cause of magma generation. Basaltic magma also forms in subduction zones. In this case, the primary cause is thought to be addition of water to hot upper asthenosphere rocks just above the top of the subducting plate. The water is driven off the wet upper layers of oceanic crust of the subducting plate as it is heated. This process may be augmented by pressure relief in circulating cells above the subducting plate.

The effect of composition on melting point can be illustrated by a simple analogy. In winter, salt is spread on icy roadways and sidewalks to lower the melting point of ice. In a similar way, the addition of water lowers the melting point of rock.

Granitic magma forms principally where basaltic magma, generated by any of the three processes described above, rises into the base of continental crust. Basaltic magma is much hotter than the melting point of the continental crust. Therefore it partially melts large volumes of lower continental crust, generating granitic magma. Thus, most granites are associated with continental crust.

After discussing the different kinds of behavior of basaltic and granitic magmas, and the ways in which magmas form, in Sections 7.1 and 7.2, we describe the different forms of intrusions and volcanic landforms in Sections 7.3 and 7.4. Section 7.5 describes the relationships among the different types of tectonic plate boundaries and igneous activity. The chapter ends with discussions of ash-flow tuffs and calderas, and the effects of volcanic activity on human settlements.

Two special topics boxes on the 1980 Mount St. Helens eruption and the relationships between volcanic eruptions and climate, augment the chapter.

Answers to Discussion Questions

1. Most batholiths are of granitic to intermediate composition, and most consist of medium- to coarse-grained rock. However, many dikes and sills are of granitic or intermediate composition and medium or coarse texture. Thus, it would not be possible to distinguish among hand samples of this type from a batholith, a dike, or a sill. However, if you were given a hand sample of basalt or andesite, and asked to guess whether it came from a batholith, dike, or sill, it would be a good guess that it came from a dike or sill, but you could not distinguish between the two.

2. If the sill cooled at depth, a coarse-grained intrusive rock would form whereas extrusive rocks are commonly fine-grained or porphyritic. The occurrence and placement of vesicles also provides clues to the environment of formation. Additionally, many lava flows have autobrecciated bases and tops; sills do not. Sills may have chilled margins and produce contact metamorphism at both contacts.

3. No. Most granite batholiths formed at depths of 5 kilometers or more, and are exposed today as a result of uplift and erosion.

4. Basalt plateaus form as great amounts of basaltic magma flood onto the Earth's surface, and thus are extrusive. Their rocks are typically fine-grained, and thus differ texturally from the typically medium- to coarse-grained batholiths. Additionally, most batholiths are of granitic to intermediate composition, and are not basaltic.

5. This question is one that has troubled geologists for decades. Most modern igneous petrologists and tectonicists feel that granitic to intermediate magma coalesces from small, dispersed blobs that form near the base of continental crust, into a large one. The large blob then rises slowly as a liquid or plastic mass by shouldering aside hot, plastic country rock. The plastic country rock then flows back to fill in behind the rising mass of granitic magma after it passes. Most such magma bodies solidify within the crust because they rise slowly, and lose water as pressure decreases as they rise.

6. The maria are basalt flows that erupted <u>after</u> the meteorite craters formed. Ask the students to describe the Moon's topography if the meteorite bombardment had occurred after the basalt flows had cooled and solidified.

7. A solution is a homogeneous mixture of two or more substances in a single phase, whereas a liquid with bubbles in it is a heterogeneous mixture of gas and liquid. Violent volcanic eruptions can occur when dissolved gases come out of solution to form bubbles as the pressure is reduced. Refer again to the analogy of the carbon dioxide in a bottle of soda or beer. Additionally, vesicles form when gas exolves from magma.

8. Volcanoes form under a variety of conditions, depending on the temperature, pressure, and composition of the interior of a planet. On

Triton, the largest of Neptune's moons, flat plains are believed to be methane lava flows. Obviously methane lava (melting point -182°C) forms under different conditions from basaltic lava, but the concepts are the same. Magma forms when material in the interior of a planet is heated or when its pressure or composition are altered. In order to learn more about the geology of other planets, you would need data on the composition of the lava, its temperature, and tectonic activity on the planet.

9. Volcanic eruptions commonly form symmetrical, conical mountains with gentle to moderate slopes; the flanks can then be steepened by weathering and erosion. However, some volcanoes are asymmetrical because the eruption is lateral. The 1980 eruption of Mount St Helens is one such example.

10. In the San Juans, some rising granitic and intermediate magma erupted as ash flow tuffs and related volcanic rocks. As the magma rose, portions were emplaced and solidified at shallow depth in the crust. Thus, both intrusive and extrusive rocks formed from the same or related magma body.

11. Mount St. Helens has erupted more frequently but less catastrophically than the Yellowstone region. If you lived very close (5 km), the high frequency of eruption would make Mount St. Helens the more dangerous location. Fifty kilometers is probably close enough to Mount St Helens to be dangerous. The blast from the 1980 eruption blew down trees 25 kilometers from the vent, and ash flows and mudflows followed river valleys for tens of kilometers. Therefore, even at 50 kilometers, the danger from Mount St. Helens is probably greater than from the Yellowstone region. However, another cataclysmic Yellowstone caldera eruption would initiate fires and eject ash that could cause damage and loss of life 500 kilometers away. Thus, while the probability that an eruption will occur is lower, there is a finite danger in living 500 kilometers from Yellowstone.

Selected Bibliography

A general text on volcanoes is:
Robert Decker and Barbara Decker: Volcanoes. New York, W.H. Freeman, 1989. 285 pp.

Magma chambers are reviewed in:
Bruce D. Marsh: Magma Chambers. Annual Review of Earth and Planetary Sciences 19 (439), 1989.

The threat that volcanoes pose to settlement in North America is discussed in:
Richard A. Kerr: Domesday Book of the World's Volcanoes. Science, 213:856, 1981.

Richard A. Kerr: Volcanoes to Keep an Eye On. Science, 221:634, 1983.

The Yellowstone Park region is discussed in:

Robert B.Smith and Robert L. Christiansen: Yellowstone Park as a Window on the Earth's Interior. Scientific American, February 1980.

Calderas are discussed in:

Peter Francis and Stephen Self: Collapsing Volcanoes. Scientific American, June 1987.

Peter Francis: Giant Volcanic Calderas. Scientific American, June 1983.

Grant Heiken, Fraser Goff, Jamie N. Gardner, W. S. Baldridge, J. B. Hulen, Dennis L. Nielson, David Vaniman: The Valles/Toledo Caldera Complex, Jemez Volcanic Field. Annual Review of Earth and Planetary Sciences 18 (27), 1990.

John B. Rundle and David P. Hill: The Geophysics of a Restless Caldera -- Long Valley, California. Annual Review of Earth and Planetary Sciences 16 (251), 1988.

The Mount St. Helens eruption is the topic of:

Robert Decker and Barbara Decker: The Eruptions of Mount St. Helens. Scientific American, 1981.

Accounts of some historical eruptions are given in:

Richard B. Stothers: The Great Tambora Eruption in 1815 and its Aftermath. Science, 224:1191, 1984.

Peter Francis and Stephen Self: The Eruption of Krakatau. Scientific American, August 1983.

An excellent article on the relationship between volcanoes and climate is:

Michael R. Rampino, Stephen Self, and Richard B. Stothers: Volcanic Winters. Annual Review of Earth and Planetary Sciences 16 (73), 1988.

Chapter 7

Plate Tectonics, Volcanoes, and Plutons

Multiple Choice:

1. Huge amounts of magma form in subduction zones because of (a) pressure-relief melting; (b) addition of water to the asthenosphere; (c) frictional heat caused by scraping plates; (d) none of the above (e) a, b, and c.

2. A drop in pressure can melt a hot rock in a process called (a) pressure-relief melting; (b) frictional heat; (c) dehydration; (d) asthenospheric oozing; (e) none of the above.

3. Parts of the upper mantle or crust melt which then solidifies to form
(a) sedimentary rock; (b) igneous rock; (c) magma; (d) mantle; (e) none of the above.

4. The magma that most commonly rises all the way to the Earth's surface to erupt from a volcano is
(a) granitic magma; (b) silica-rich magma; (c) basaltic magma; (d) plutonic magma.

5. Granitic magma is highly viscous because
(a) of its high water content; (b) it is mafic; (c) its long silica chains become entangled; (d) of its high melting point; (e) all of the above

6. Most basaltic magma erupts at the Earth's surface because
(a) it contains little water; (b) it contains huge amounts of water; (c) it is more viscous than granitic magma; (e) it has a high silica content.

7. Loss of water from a rising granitic magma causes it to
(a) erupt at the surface; (b) solidify within the crust; (c) remain liquid until it reaches the Earth's surface; (d) none of the above.

8. An intrusive mass with an outcrop area of over 100 square kilometers is a
(a) volcanic neck; (b) batholith; (c) stock; (d) dike; (e) volcano.

9. The immense quantities of magma required to form granite batholiths are most often found at
(a) spreading centers; (b) subduction zones; (c) the Mid-Atlantic ridge; (d) transform faults

10. As lava cools and solidifies from the surface down, cracks grow downward to hotter zones where the last bit of magma is solidifying. Such cracks are called
(a) batholiths; (b) sills; (c) columnar joints; (d) stocks; (e) plutons.

11. Lava with numerous holes that give it a jagged, rubbly surface is
(a) pahoehoe; (b) aa; (c) granitic; (d) plutonic; (e) cosmic.

12. Pillow lava
(a) erupts under water; (b) forms from spheroidal weathering; (c) usually has a granitic composition; (d) results from shrinkage of lava domes; (e) is rare in oceanic crust.

13. Material heaved from volcanoes include
(a) bombs; (b) cinders; (c) ash; (d) spindles; (e) all of these.

14. Pyroclastic rocks form
(a) from pahoehoe flows; (b) during explosive eruptions; (c) during intrusion of sills; (d) as batholiths force country rocks aside; (e) when igneous rocks weather into sediments.

15. The gentlest, least catastrophic type of volcanic eruption occurs when magma is so fluid that it simply oozes from cracks to form
(a) flood basalts; (b) caldera eruptions; (c) cinder cones; (d) shield volcanoes; (e) nuee ardente.

16. If basaltic magma is too viscous to form a lava plateau, but still quite fluid, it heaps up slightly forming
(a) flood basalts; (b) pillow lava; (c) submarine volcanoes; (d) shield volcanoes; (e) nuee ardente.

17. The "ring of fire", a zone of concentrated volcanic activity encircling the Pacific Ocean basin, is located directly over
(a) subduction zones; (b) rift zones; (c) the mid-oceanic ridge; (d) a lava plateau; (e) the San Andreas fault.

18. The island of Hawaii is an example of
(a) igneous activity above a rift zone; (b) igneous activity above a mantle plume; (c) igneous activity above a subduction zone; (d) a batholith; (e) a volcanic neck.

19. Which magma type has the most violent eruptions?
(a) basaltic; (b) Hawaiian; (c) Icelandic; (d) granitic.

20. When dry granitic magmas reach the Earth's surface, gas bubbles in the uppermost layer of magma create
(a) a highly pressurized, frothy, expanding mixture of gas and

liquid magma; (b) shield volcanoes; (c) lava plateaus; (d) composite cones; (e) pillow basalts.

21. When an ash flow comes to a stop, most of the gas escapes into the atmosphere, leaving behind a chaotic mixture of volcanic ash and rock fragments called
(a) ash-flow basalt; (b) pillow lava; (c) a submarine volcanoes; (d) a shield volcanoes; (e) ash-flow tuff.

22. Calderas form
(a) from gentle flows of low-viscosity lava; (b) from dry magma with no gases; (c) from granitic magma; (d) when gases in the country rock dissolve the magma; (e) only under the ocean.

23. Yellowstone Park is an example of three overlapping
(a) shield volcanoes; (b) mantle plumes; (c) calderas; (d) lava plateaus; (e) composite cones.

24. Hazards to humans from volcanic eruptions include
(a) pyroclastic flows; (b) mud flows; (c) post eruption starvation; (d) tsunamis; (e) all of the above.

True or False:

1. The asthenosphere is so hot that if water is added to it, it melts.

2. Granitic crust forms at the mid-oceanic ridge.

3. Granitic magma has longer chains of silicate tetrahedra than basaltic magma.

4. Basaltic magma has shorter silica chains and is therefore more viscous and does not flow easily.

5. If rising magma contains little water, the dropping pressure can override the cooling effect, and the magma can remain liquid and rise to erupt at the Earth's surface.

6. Basalt volcanoes are the most common.

7. A columnar joint is a tabular, or sheet-like, intrusive igneous rock that forms when magma oozes into a fracture in country rock.

8. A sill forms when magma oozes between layers of country rock.

9. Aa lava cools and stiffens as it flows, forming basalt with smooth, glassy-surfaced, wrinkled, or "ropy" ridges.

10. The gentlest, least catastrophic type of volcanic eruption occurs when magma is so fluid that it simply oozes from cracks as nuee ardente.

11. A cinder cone is usually active for only a short period of time because once the gas escapes, the driving force behind the eruption is removed.

12. Volcanic pipes are conduits filled with the last bit of magma that solidified within it.

13. Continental rifting commonly forms both basalt volcanoes and granite plutons.

14. The granitic magmas that do rise to the surface and explode probably start out with more water than "normal" granitic magma.

15. Magma that explodes to form ash-flow tuffs and calderas commonly erupts only once.

16. The periodicity of Yellowstone eruptions, the presence of shallow magma, and the well-known tendency of magma of this type to erupt multiple times all suggest that a fourth eruption may be due.

17. When Mount St. Helens erupted in 1980, it exploded with the force of 500 atomic bombs of the size used on Hiroshima in World War II.

Completion:

1. Heat, addition of water, and _______ ______ cause rocks to melt and form magma.

2. A ______ ______ is a volcanic region at the Earth's surface directly above a rising plume of hot, plastic mantle rock.

3. Long chains of silica tetrahedra form in magma with a ______ composition.

4. Granitic magma contains up to _____ percent water.

5. When magma rises into the crust and solidifies it forms a ______.

6. Fluid lava forms ______ ______ and shield volcanoes.

7. Explosively erupted rock particles or magma are called ______ rock.

8. When blobs of molten lava spin through the air, they solidify to form ______ ______.

9. A ______ ______ is a small volcano, as high as 400 meters, made up of pyroclastic fragments blasted out of a central vent at high velocity.

10. Mount St Helens and Mount Rainier are examples of ______ ______.

11. About 70 percent of the Earth's volcanic activity occurs along a circle of subduction zones in the Pacific Ocean called the ______ ______ ______.

12. Kimberlite volcanic pipes are important because a small amount of the carbon in them crystallized to form ______.

13. A volcanic rock that is so full of gas bubbles that it floats is called _____.

14. The french term, nuee ardente, or "glowing cloud" describes a/an ______ ______.

15. A tough, hard rock that forms when a hot ash flow solidifies is ______ ______.

16. The circular depression left when the roof of a magma chamber collapses is called a/an ______.

17. When Mount Krakatoa erupted on an uninhabited island, a/an ______ caused the death of about 35,00 people on nearby islands.

Answers for Chapter 7

Multiple Choice: 1. e; 2. a; 3. b; 4. c; 5. c; 6. a; 7. b; 8. b; 9. b; 10. c; 11. b; 12. a; 13. e; 14. b; 15. a; 16. d; 17. a; 18. b; 19. d; 20. a; 21. e; 22. c; 23. c; 24. e

True or False: 1. T; 2. F; 3. T; 4. F; 5. T; 6. T; 7. F; 8. T; 9. F; 10. F; 11. T; 12. T; 13. T; 14. F; 15. F; 16. T; 17. T

Completion: 1. pressure-relief melting; 2. hot spot; 3. granitic; 4. 15; 5. batholith; 6. lava plateaus; 7. pyroclastic; 8. volcanic bombs; 9. cinder cone; 10. composite cones; 11. ring of fire; 12. diamond; 13. pumice; 14. ash flow; 15. welded tuff; 16. caldera; 17. tsunami

CHAPTER 8

Mountains and Geologic Structures

Discussion

In this chapter we introduce the fundamental concepts of structural geology. These concepts are presented in the context of the processes that form most geological structures: The development of orogens, or mountain building. The plate tectonics concepts discussed in the previous chapters are applied to the building of mountain ranges.

Structural geology is a fascinating and informative branch of geology. We feel that the important structural concepts for an introductory course are few and simple: Plate movements generate tectonic stress, particularly near plate boundaries. The stress, in turn, causes elastic, plastic, and/or brittle deformation in rocks. Deformation occurs as folds, faults, and joints. Relationships among types of plate boundaries, types of stress, and geologic structures are described.

Intuition leads students, and many of us geologists, to conclude that a simple relationship must exist among each of the three types of plate boundaries and the type of tectonic stress that predominate at each: rift boundaries seem to logically involve extensional stress; transform boundaries should produce shear stress; and collisional boundaries should involve compressive stress. We point out that this is commonly the case, particularly for rift and transform boundaries. However, collisional, or subduction boundaries often are accompanied by regional extension. As Warren Hamilton frequently observes, much of western South America seems currently to be under extension, and simultaneously is a subduction margin.

After introducing geological structures and the general idea that mountain chains form at tectonic plate boundaries, we use these ideas to describe the origins of two of the Earth's greatest mountain chains: the Andes and the Himalayas. Nearly all students have some geographic familiarity with both mountain chains. Their respective origins and the fact that both mountain chains are still in the process of rising today are intrinsically interesting to most students. Further, reading about the geological processes that formed these mountain chains strongly reinforces students' understanding of the processes themselves.

Answers to Discussion Questions

1. Most continental mountain chains form at convergent plate boundaries either where subduction of oceanic lithosphere occurs at a continental margin (the Andes along the west coast of South America), or where two continents collide (the Himalayas). Both environments involve the generation of huge quantities of magma. In the case of continent-continent collisions, the magma forms before the collision, when an Andean margin precedes the continents, themselves, collide. Both environments also involve deformation of vast

regions near the respective plate boundaries. As a result of deformation and the addition of magma to continental crust, immense mountain chains such as the Andes and the Himalayas rise near these plate boundaries.

Continental rifting results from development of a divergent plate boundary in continental crust. High escarpments result from normal faulting, and volcanic mountains form in such environments, as in the East African Rift. Generally, little folding or deformation other than normal faulting occur in continental rifts because the stress is mainly extensional.

A transform plate boundary exists along the San Andreas fault in western California. Here, small, folded and faulted mountain ranges are rising along the fault zone. Somewhat similar ranges are forming along the great strike-slip faults in Tibet, China, and Mongolia, north of the Himalayas. Although this region is not a transform plate boundary, the huge strike-slip faults generate similar tectonic forces and structures.

2. Divergent plate boundaries, whether in oceanic or continental lithosphere, are characterized by extension, normal faulting, grabens, and basaltic volcanism.

The dominant feature of a transform plate boundary is a system of huge, lithosphere-deep, strike-slip faults. Secondary stress generated along these faults creates both compressive and extensional features in relatively small crustal blocks adjacent to the main faults. Basaltic volcanism may occur if the transform is "leaky".

All three types of stress, compression, extension, and shear, may occur at convergent boundaries, at different times or in different places at the same time. Compressional stress can be dominant where an island arc docks against a continental margin, or where two continents collide. It may also occur at a continental margin when convergence rates are high. Compression occurs within the subduction complex adjacent a trench. Extension can also accompany plate convergence, as seems to be the case currently in parts of the Andes. If subduction occurs because of relatively high density of a cool, old, portion of a plate, little or no compression should be expected in the obducting plate except in the vicinity of the subduction complex, and extension is likely. Shearing occurs along oblique subduction zones. Horizontal shearing certainly occurs in subduction complexes, as a result of compression.

3. The most common explanation for development of forearc basins is that they are depressions in oceanic crust bounded by a magmatic arc on one side, and a rising subduction complex on the trench side of a subduction zone. The subduction complex rises due to underthrusting of oceanic lithosphere. Back arc basins probably form due to loading of the lithosphere by addition of mass in the magmatic arc. This additional mass isostatically depresses the lithosphere on both sides of the arc, forming a back arc basin, and enhancing the forearc basin.

A back arc basin in continental crust should contain large amounts of continental clastics as well as clastics derived from the magmatic arc. This environment also seems to favor carbonate deposition. A forearc basin should contain a higher proportion of clastics derived from the magmatic arc, since it may be cut off from the source of non-magmatic continental clastics by the magmatic arc. Carbonates also form in forearc basins. In an island arc, where a large mass of continental crust is not involved, sediment in forearc and back arc basins may be more similar to each other.

4. Rocks of subduction complexes are highly varied, which is probably why "complex" is part of the name. They include rocks from all three layers of oceanic crust as well as upper mantle, which have been sheared off the top of the subducting slab. Thus, they consist of oceanic sediment including deep sea mud and bedded chert; pillow, sheeted dike, and other types of basalts; gabbro; and upper mantle peridotite and related rocks. They also include sediment derived from the magmatic arc and continental side of the trench. These rock types include turbidites and other clastic units that flowed into the trench. Carbonates and other rocks from a forearc basin may also be present.

Because of the intense tectonic activity associated with subduction, the rocks tend to be intensely sheared and folded. In some cases, they have been dragged downward to considerable depth before returning to the Earth's surface. Because the down-going slab tends to remain cool relative to depth, high-pressure but relatively low-temperature metamorphism of the blueschist and related metamorphic facies are common.

5. Early development of the Himalayas must have been similar to the present stage of development of the Andes. Major differences developed later, when India collided with southern Asia. That collision transformed the Himalayan region from an Andean-type margin to a continental collision plate margin. Thus, 60 million years ago the Himalayan region may have been very similar to the modern Andes.

6. Large masses of granitic rocks are found on the Tibetan side of the Himalayas. They are located on the northern side of the continental suture and the early subduction zones that preceded the continental collision. The southern side of the suture is characterized by folded and faulted sedimentary rocks. The igneous rocks concentrate on the northern side of the suture because subduction that preceded the continental collision involved oceanic lithosphere diving northward beneath the southern margin of Tibet.

7. The answer to this question is easiest to explain by forming a tight fold with a piece of paper, tilting the fold to simulate a plunging fold, and then using scissors to simulate erosion creating a level portion of the Earth's surface. The resulting pattern in the fold is V-shaped.

8. All planets in the Solar System were bombarded by swarms of meteorites during the first half-billion years. Thus, initially all planets were pock-marked by meteorite impact craters. Any later tectonic activity would produce features that would cross-cut those craters. Features that would indicate tectonic activity include mountain ranges, rift valleys, lava flows, trenches (like oceanic trenches on Earth), and volcanic cones. It would be necessary to distinguish features formed by tectonic activity from those created by weathering and erosion, which also would cross-cut impact craters.

Selected Bibliography

P. Molnar: The Geologic History and Structure of the Himalaya. American Scientist, 74:144-154, 1986.

Z. Peizhen, P. Molnar, B. Burchfiel, and L. Royden: Bounds on the Holocene Slip Rate of the Haiyuan Fault, North-Central China. Quaternary Research, 30: 151-164, 1988.

F. Herve, E. Godoy, M. Parada, V. Ramos, C. Rapela, C. Mpodozis, and J. Davidson: A General View on the Chilean-Argentine Andes, with Emphasis on Their Early History. Geodynamics Series. Washington, D. C., American Geophysical Union, 1987.

B. Coira, J. Davidson, C. Mpodozis, and V. Ramos: Tectonic and Magmatic Evolution of the Andes of Northern Argentina and Chile. Earth Science Reviews, 18: 303-352, 1982.

K. Berg, C. Breitkreuz, K. Damm, S. Pichowiak, and W. Zell: The North Chilean Coast Range - An Example for the Development of an Active Continental Margin. Geologische Rundschau, 72: 715-731, 1983.

J. Suppe: Principles of Structural Geology. Englewood Cliffs, N. J., Prentice-Hall, Inc., 1985. 560 pp.

Chapter 8

Mountains and Geologic Structures

Multiple Choice:

1. A series of mountains or mountain ridges that are closely grouped and similar in age and mode of formation is called
(a) a range; (b) a rift valley; (c) a subduction zone; (d) fault zone.

2. Mountains nearly always occur in ranges and chains because
(a) the forces that create them operate over large, linear regions of the Earth's crust; (b) the forces that create them operate in small, isolated localities; (c) they are not closely grouped in age or mode of formaticn; (d) none of these.

3. Nearly all of the Earth's mountains and mountain ranges are found at
(a) the interior of plates; (b) plate boundaries; (c) small isolated localities; (d) places where there is no tectonic activity; (e) at places where the elevation is less than 300 meters.

4. Mountain building is associated with
(a) volcanic eruptions; (b) intrusion of plutons; (c) earthquakes; (d) folding and faulting of rocks; (e) all of these.

5. Most of the Earth's great mountain ranges, with the exception of the mid-oceanic ridge, grew at
(a) calderas; (b) island arcs; (c) divergent plate boundaries (d) convergent plate boundaries; (e) none of these.

6. Rocks in mountainous areas at plate boundaries are likely to be
(a) smooth and unbroken; (b) intensely deformed; (c) very young; (d) only sedimentary.

7. A fracture along which rock on one side has moved relative to rock on the other side is a
(a) joint; (b) fault; (c) fold; (d) none of these.

8. An anticline is a fold that arches
(a) downward; (b) up and down; (c) upward; (d) across an oceanic trench; (e) under a monocline.

9. The backbone of a fold, where two limbs meet is
(a) an anticline; (b) the axis; (c) the axial plane; (d) the plunge; (e) a limb line.

10. Slip is
(a) the distance that rocks on opposite side of a fault have moved; (b) The distance that rocks on the same side of a fault have moved; (c) sometimes hundreds of meters at large faults; (d) a and c; (e) none of these.

11. Many faults move repeatedly because
(a) tectonic stress commonly continues to be active in the same place over long periods of time; (b) once a fault forms, it is easier for movement to occur again along the same fracture than for a new fracture to develop nearby; (c) a and b; (d) tectonic stress never occurs in the same place over long periods of time; (e) once the stress is removed it is not likely to rebuild.

12. A fault where the hanging wall moves down relative to the footwall is
(a) a thrust fault; (b) a reverse fault; (c) a normal fault; (d) a joint; (e) an overthrust fault.

13. Normal faults, grabens, and horsts are common
(a) where the crust is being pulled apart; (b) at convergent boundaries; (c) at hot spots; (d) where plates collide; (e) in continental cratons.

14. Which type of geologic structure accommodates crustal shortening?
(a) reverse faults; (b) normal faults; (c) strike-slip faults; (d) thrust faults.

15. The San Andreas fault is an example of a
(a) reverse fault; (b) normal fault; (c) strike-slip fault; (d) thrust fault; (e) tight fold.

16. A joint is a fracture
(a) in which rock on one side has moved relative to rock on the other side; (b) in which the rocks on each side have not moved; (c) that occurs only in sedimentary rock; (d) where there has been over 10 kilometers of displacement.

17. Where two oceanic plates converge, one subducts beneath the other and dives into the mantle forming
(a) a mid-ocean ridge; (b) a strike-slip fault; (c) an oceanic trench; (d) reverse fault; (e) a thrust fault.

18. India collided with Asia to form the
(a) Alps; (b) Himalayas; (c) Andes; (d) African rift; (e) Aleutian arc.

19. The Andes Mountains were formed by
(a) subduction; (b) continent-continent collision; (c) a shield

volcano; (d) a hot spot; (e) a transform plate boundary.

20. When two continents collide, they weld into a single mass of continental crust along a
(a) transform fault; (b) hanging wall; (c) suture zone; (d) rift; (e) Pangaea.

True or False:

1. Rocks trapped between colliding plates bend and fracture.

2. If you were to walk along the top of a horizontal anticline, you would be walking along a level fold axis.

3. A fold usually reflects compressional tectonic stress.

4. If tectonic force fractures the Earth's crust, rocks on opposite sides of the fracture may move past each other to create a fault.

5. Stored elastic energy does not generate earthquakes.

6. If tectonic forces stretch the crust over a large area, many thrust faults may develop, allowing numerous grabens to settle downward along the faults.

7. The blocks of rock between the down dropped grabens then appear to have moved downward relative to the grabens.

8. At most convergent plate boundaries, where two plates collide, compressional forces squeeze rocks, forming folds and reverse and thrust faults.

9. The western Aleutian Islands and most of the island chains of the southwestern Pacific basin are examples of forearc basins.

10. The modern Himalayas continue to grow as a result of a collision between a tectonic plate carrying oceanic crust and another carrying continental crust.

11. The Andes developed from a collision between a tectonic plate carrying oceanic crust and another carrying continental crust.

12. Subduction at a continental margin forms an Andean margin.

13. Before it collided with Asia, India drifted north at rates up to 20 cm/yr.

14. During the formation of the Himalayas, an Andean type margin must first develop as oceanic crust separating Asia and India subducted.

15. The Himalayas are sinking as much as 1 centimeter each year.

Completion:

1. ______ ______ builds mountains.

2. A/an ______ ______ is any feature produced by deformation of a rock.

3. A/an ______ is a special kind of fold with only one limb.

4. A circular or elliptical anticline resembling an inverted bowl is called a/an ______.

5. In a reverse fault the hanging wall has moved ______ relative to the footwall.

6. A/an ______ ______ boundary is an immense strike-slip fault that cuts through the entire lithosphere.

7. Slices of basalt from oceanic crust and pieces of the upper mantle, and are then scraped off and mixed in with the sea-floor sediment which are highly deformed, sheared, faulted, and metamorphosed are collectively termed a/an ______ ______.

8. A subduction complex grows by addition of the newest slices at the ______ of the complex.

9. A/an ______ ______ is a volcanic mountain chain plus associated sedimentary and metamorphic rocks.

10. A/an ______ ______ is a depression between a subduction complex and an adjacent mountain range.

11. The Andes Mountains are formed by ______ at a continental margin.

12. By about 50 million years ago, all of the oceanic lithosphere between India and Asia had been consumed by ______.

13. A ______ ______ is responsible for the formation of the Himalayas.

14. The leading edge of India began to slide under Tibet in a process called ______.

Answers for Chapter 8

Multiple Choice: 1. a; 2. a; 3. b; 4. e; 5. d; 6. b; 7. b; 8. c; 9. b; 10. d; 11. c; 12. c; 13. a; 14. a; 15. c; 16. e; 17. c; 18. b; 19. a; 20. c;

True of False: 1. T; 2. T; 3. T; 4. T; 5. F; 6. F; 7. F; 8. T; 9. F; 10. F; 11. T; 12. T; 13. T; 14. T; 15. F

Completion: 1. Tectonic activity ; 2. geologic structure; 3. monocline; 4. dome; 5. upward; 6. transform plate; 7. subduction complex; 8. bottom or seaward edge; 9. island arc; 10. forearc basin; 11. subduction; 12. subduction; 13. continent-continent collision; 14. underthrusting

CHAPTER 9

Geological Evolution of North America

Discussion

This chapter uses the geological concepts taught in the first 8 chapters to describe and discuss the geological evolution of North America.

The geology and evolution of our continent are ignored by other introductory texts. We have included the subject for three reasons. First, most beginning students are interested in understanding the geology of specific and familiar places. By this point in an introductory course, students have learned enough geology to use it. This chapter allows them to apply their newly-learned skills toward a study of their own place in the world. In the course of discussing the evolution of our continent, we describe the development and history of each portion of North America. As a result, a student anywhere on the continent can relate the information to the areas around his or her home and college or university. The information also helps to place field trips and lecture discussions of local geology in a larger geological framework.

The second reason for including a chapter on North America is pedagogical. We feel that general concepts and principles of any kind are most successfully learned when they are applied. This is particularly true when they are applied to something that is already interesting and well-known in other ways. All college and university students are familiar with the general geography of North America, and they are certainly familiar with their own states and regions. Thus, using concepts learned from earlier chapters to describe the geological evolution of our continent should help make geological thinking a permanent part of students' views of the world.

Third, recent advances in plate tectonics ideas have allowed geologists to assemble a relatively simple, logical, and attractive story of the evolution of North America. When told as an unfolding story, it is easily and eagerly understood and assimilated by introductory-level students.

A special topics box describes the relationships between geology and radioactive waste storage in the United States.

Answers to Discussion Questions

1. The Arcasta gneiss near Great Slave Lake, Northwest Territories, Canada, give ages of 3.96 billion years. The fact that they are metamorphic rocks means that some earlier rock must have existed, and was metamorphosed. Therefore, 3.96 billion years is a minimum age for the North American craton.

2. This question has no real answer. Instead, it is an invitation for students to discuss the differences between data, or facts as they are known, and theory, or the interpretation of data. In this chapter, we attempt to distinguish clearly between data regarding the geology of our continent, and

the models for the evolution of the continent. Thus, this chapter provides a good opportunity for a professor to clarify the critical distinction in all sciences between data and theory.

3. These models are described in the chapter, and in greater detail in Hoffman's papers listed in the Selected Bibliography. A professor may wish to suggest that interested students read one or more of those papers.

4. Microcontinents may have formed as island arcs, or collections of island arcs accreted at a subduction zone. Some of the oldest microcontinents might be the remains of primordial crust.

5. Supercontinents probably form because oceanic crust is rapidly consumed by subduction, whereas continents accrete. When subduction has gone on for a long enough time, all continental crust accretes together to form a supercontinent.

Another way to answer the question is to observe that the Earth is a spheroid. When a supercontinent breaks up, the fragments spread outward under its own weight. If they keep spreading outward, eventually, they come together again on the opposite side of the spheroid.

6. Granites of this type are sometimes called "anorogenic granites," referring to the fact that they seem to form without major folding of associated rocks or mountain building. As supercontinent rifting begins, mantle rock must rise beneath the rift zone. The decreasing pressure causes pressure relief melting, producing large quantities of basaltic magma. Initially this basaltic magma would rise into the base of the supercontinent, in turn melting the lower continental crust. This melting would generate large amounts of granitic magma.

7. The San Andreas Fault began to develop as the western edge of North America reached and overrode the East Pacific Rise. Thus, it resulted from a major change in relative plate motions along the western margin of the continent. That change most importantly involved a transition from plate convergence to transform motion. Thus the plate boundary changed from a subduction boundary to a transform plate boundary. According to some models, particularly that of T. Atwater, frictional drag along the San Andreas Fault created extensional forces east of the fault, causing normal faulting of the Basin and Range Province. The same forces must have affected the region of the Colorado Plateau, but for some reason it simply rotated clockwise rather that undergoing Basin and Range type normal faulting. Some geologists have suggested that the difference in behavior was due to thicker, stronger crust beneath the Colorado Plateau.

8. Figure 9-1 can be used for this question. Extrapolation of future plate movement can be done using a combination of plate movement directions and rates. Generally, California west of the San Andreas fault will slide northward toward the Aleutian trench, the Pacific will grow smaller, and the Atlantic will grow larger. Many other changes can be speculated upon based on current plate motions. It is entertaining to further speculate on effects on

climate, human communities, and orogenies. It is also interesting to speculate about changes in current plate boundaries, and development of new ones. For example, what if the East African Rift becomes a major oceanic spreading center? The Rio Grande Rift? The Snake River Plain? Where would subduction begin to accommodate new spreading centers?

Selected Bibliography

Paul F. Hoffman: United Plates of America, the Birth of a Craton: Early Proterozoic Assembly and Growth of Laurentia. Annual Review of Earth and Planetary Sciences 16 (543), 1988.

C. Stearn, R. Carroll, and T. Clark: Geological Evolution of North America. New York, John Wiley & Sons, 1979. 566 pp.

R. Redfern: The Making of a Continent. New York, Times Books, 1983. This book accompanies a good PBS television series on the origin of North America. Video tapes of the series are available.

P. Hoffman: Speculations on Laurentia's First Gigayear (2.0 to 1.0 Ga). Geology, 17: 135-138.

A. Bally and A. Palmer, Eds.: Geology of North America- An Overview. Boulder, Co., Geological Society of America, 1989.

Radioactive waste disposal, discussed in the Earth Science and the Environment box, is reviewed in:
Stephen Brocoum, Kathleen Mihm, Michael Cline, and Scott Van Camp: Department of Energy. Geotimes, (10), January 1989.

Carl A. Johnson: Nevada Nuclear Waste Product Office. Geotimes, (12), January 1989.

Phillip S. Justus and Newton K. Stablein: U.S. Nuclear Regulatory Commission. Geotimes, (14), January 1989.

Konrad B. Krauskopf: Geology of High-Level Nuclear Waste Disposal. Annual Review of Earth and Planetary Sciences 16 (173), 1988.

Charles R. Malone: The Yucca Mountain Project. Environmental Science and Technology, 23:12, (1452), 1989.

Konrad B. Krauskopf: Radioactive Waste Disposal and Geology. London, Chapman and Hall, 1991, 145 pp.

Chapter 9

The Geological Evolution of North America

Multiple Choice:

1. The oldest rocks in North America are ____ years old.
(a) 5 billion; (b) nearly 4 billion; (c) 2 billion; (d) 4 million; (e) 10,000.

2. The interior part of North America is
(a) the craton; (b) the shelf; (c) the basin; (d) the orogenic lands; (e) the plate boundary.

3. The mountain chains of North America are ______ the craton.
(a) older than; (b) younger than; (c) the same age as; (d) none of these.

4. If you could strip away the relatively young sedimentary rocks from the North American craton you would find
(a) younger igneous rocks: (b) mountain chains; (c) older igneous and metamorphic rocks; (d) continental shelves; (e) coastal plains.

5. According to the supercontinent model
(a) rifting causes the fragments to spread apart; (b) rifted continental fragments migrate halfway around the globe and then collide on the far side to reassemble as a new supercontinent; (c) the breakup of one supercontinent leads to the assembly of a new one; (d) a, b, and c; (e) all the geologic provinces have the same radiometric age.

6. ______ ______ swept all of the island arcs, microcontinents, and other masses of continental crust on the Earth's surface together into a single great landmass called Pangaea I.
(a) Tectonic movement; (b) earthquakes; (c) mountain building; (d) hot spots

7. The first supercontinent formed by collisions of microcontinents and island arcs about ____ years ago.
(a) 4.5 billion; (b) 3.2 billion; (c) 2 to 1.8 billion; (d) 5 million; (e) 600,000.

8. When Pangaea I broke apart,
(a) the provinces of the North American craton separated from each other; (b) the provinces of the North America craton remained welded together; (c) the fracture led to the formation of the Appalachians; (d) the fracture lead to the formation of the Cordilleran.

9. After Pangaea I split up, the separate fragments of continental crust migrated across the Earth and then reassembled, forming a second supercontinent called Pangaea II, ______ years ago.
(a) 4.5 billion; (b) 3.2 billion; (c) 60,000; (d) 5 million; (e) about 1.0 billion

10. As Pangaea II broke up, the ______ portion rifted away leaving a shoreline at the western margin of North America.
(a) European; (b) African; (c) Australian; (d) Siberian; (e) Cuban

11. As Pangaea II broke up, the rift that opened on the eastern margin became an early version of
(a) Pangaea III; (b) the Atlantic Ocean; (c) Siberia; (d) the Pacific Ocean.

12. The suturing of continents caused when the North American plate reversed direction and began colliding with Africa and Europe created
(a) Pangaea III; (b) Pangaea II; (c) Pangaea I; (d) the Appalachian mountains; (e) a and d.

13. Formation of the Appalachians resembles the two-step formation of the modern
(a) Atlantic margin; (b) Andes; (c) Rocky Mountains;(d) Himalayas; (e) platform.

14. Appalachian rocks are concealed under younger coastal plain sedimentary rocks in
(a) California; (b) Utah; (c) Texas; (d) Greenland; (e) Siberia.

15. When Pangaea II broke apart sea level rose, and the low central part of the craton
(a) was alternately flooded and drained; (b) remained above sea level; (c) was covered with layers of sand, limestone and shale; (d) a and c; (e) none of these.

16. Sea-floor spreading can flood continents by
(a) displacing sea water; (b) creating more deep ocean basins; (c) creating great earthquakes; (d) producing continental platforms; (e) all of these.

17. When Pangaea III broke up and the Atlantic Ocean began to open, subduction began along the west coast of North America to form
(a) the Cordilleran mountain chain; (b) the Appalachian mountain chain; (c) the Innuitian mountain chain; (d) the Marathon mountain chain.

18. What has been added to North America by accretion in the past 200 million years?
(a) most of Alaska; (b) most of British Columbia; (c) most of California; (d) most of Washington; (e) most of all of these.

19. The added weight of the Cordilleran chain and high global sea level created ______ to the east of the mountains
(a) the Colorado plateau; (b) a swampy plain; (c) the Cascade volcanoes; (d) the accreted terranes; (e) the Rocky Mountains

20. Modern active subduction volcanoes are found in the
(a) Sierra Nevada; (b) Craton; (c) Cascades; (d) Appalachians; (e) Hawaiian Islands.

21. The Basin and Range province includes large areas of
(a) Nevada; (b) New York; (c) Tennessee; (d) southeastern Texas; (e) Kansas and Nebraska.

22. The Pleistocene ice age began ____ years ago.
(a) 20 million; (b) 2.0 million; (c) 500,000; (d) 16,000; (e) 10,000

True or False:

1. Although some of the mountains of North America are hundreds of millions of years old, they are young relative to the craton.

2. Geologic relationships among rocks within each province are continuous, whereas relationships across province boundaries are discontinuous.

3. The rocks of each province give the same radiometric ages as those of adjacent provinces.

4. The breakup of one supercontinent does not lead to the assembly of a new one.

5. The geologic differences among the provinces of the North American craton exist because each province was isolated from the others when it formed.

6. By Middle Ordovician time, North America was moving away from Europe and Africa.

7. Igneous and metamorphic rocks of the Appalachians poke up through the sedimentary rocks in the Ouachita and Marathon mountains in Arkansas, Oklahoma, and east Texas.

8. The new mid-oceanic ridge system that raised sea level when

Pangaea II broke up flooded the low central portion of North America with shallow seas.

9. Western North America has always remained the same size.

10. The igneous activity and volcanoes of the modern Cascade Range result from subduction along this part of the west coast.

11. The Colorado Plateau was greatly affected by faulting and igneous activity throughout its history.

12. Beginning about 30 million years ago, the western margin of North America reached the East Pacific rise, and the San Andreas fault formed.

13. Twenty million years ago, large continental ice caps up to 3000 meters thick formed in both the Northern and Southern Hemispheres and rapidly spread outward.

14. Five million years ago, the precursors of modern humans evolved from a branch of apes, probably in eastern Africa.

15. Mammoths, bison, caribou, moose, musk ox, and mountain sheep migrated to this continent within the last 150,000 years.

Completion:

1. The part of the craton covered by a veneer of younger sedimentary rocks is the ______.

2. Old metamorphic rocks are exposed on the part of the craton called the _____.

3. A/an ______ ______ is a large region in which rocks share a similar geologic history.

4. A/an ______ is a small mass of continental crust surrounded by oceanic crust.

5. The building of the Appalachian mountains happened during the assembly of Pangaea ______.

6. The Appalachian mountains formed from a/an ______ collision,

7. Shale and limestone deposited in shallow water as sea level rose and fell created ______ ______ ______.

8. Subduction in the Cordillera coincided with opening of the _____ Ocean.

9. The process of accretion of island arcs onto a continental margin is called _____.

10. As terranes crashed into the continent, they compressed rocks near the collision zone, forming a region called the Cordilleran _____ _____ _____.

11. Island arcs and microcontinents added to North America are _____ terranes.

12. The rocks formed from swampy plains of the Jurassic and Cretaceous periods contain numerous _____ fossils.

13. The _____ _____ fault developed to accommodate the small difference in direction between the North American and Pacific plates as North America moved westward and the Pacific plate moved northwest.

14. A region in western US characterized by alternate parallel ridges and valleys formed by crustal extension is the _____ _____ _____ province.

15. The low-lying, swampy bridge that connected North America and Asia during the Pleistocene is called _____.

Answers for Chapter 9

Multiple Choice: 1. b; 2. a; 3. b; 4. c; 5. d; 6. a; 7. c; 8. b; 9. e; 10. d; 11. b; 12. e; 13. d; 14. c; 15. d; 16. a; 17. a; 18. e; 19. b; 20. c; 21. a; 22. b

True or False: 1. T; 2. T; 3. F; 4. F; 5. T; 6. F; 7. T; 8. T; 9. F; 10. T; 11. T; 12. F; 13. T; 14. T; 15. T

Completion: 1. platform; 2. shield; 3. geologic province; 4. microcontinent; 5. III; 6. continent-continent; 7. platform sedimentary rocks; 8. Atlantic; 9. docking; 10. fold and thrust belt; 11. accreted; 12. dinosaur; 13. San Andreas; 14. basin and range; 15. Beringia

CHAPTER 10

Weathering, Soil, and Erosion

Discussion

Chapters 10 through 13 discuss surface processes, those processes that act upon the Earth's surface to sculpt and reshape landforms originally created by tectonic processes. In Chapter 10, we discuss processes and effects of weathering, soil formation and soil types, and erosion.

Chemical weathering not only affects rocks, but also man-made structures and industrial materials. Unpolluted moist air is corrosive by itself. Acid rain produced by air pollutants is even more corrosive and causes billions of dollars worth of damage each year. Many steel buildings and bridges in the United States are deteriorating. About half a century after it was built, portions of the elevated West Side Highway in New York City rusted through and collapsed, sending vehicles plummeting onto the roadway below. In 1988, New York's Williamsburg bridge was closed due to corrosion by the salt that is spread on the roadway every winter.

Chapter 22 discusses geological resources. Since there is often insufficient time to include this subject at the end of a course, some instructors may chose to incorporate many of the topics into earlier chapters. If this strategy is chosen, you may wish to discuss weathering processes that form ore deposits here. This topic is a sub-section of Section 22.6, How Ore Deposits Form.

Four kinds of agents or processes erode soil and bedrock: mass wasting, streams and other flowing water, glaciers, and wind. In this chapter we discuss erosion in general, and mass wasting specifically. Streams, glaciers, and wind are discussed in following chapters.

Mass wasting accompanies many other geological processes. We have already discussed landslides and debris flows that are initiated by earthquakes and volcanic eruptions. Valleys widen as a result of a combination of erosion by streams and glaciers and mass wasting. Sinkholes form when caverns collapse, and mass wasting is common along sea coasts as waves undercut cliffs.

An obvious similarity exists between mass wasting and avalanches. A powder avalanche is analogous to flow; the snow crystals move independently of one another. A slab avalanche is analogous to a slide because it occurs when consolidated chunks of snow break away and begin to move as coherent units. Hanging glaciers can fracture to release free-falling ice avalanches.

In both avalanches and mass wasting, it is relatively easy to identify areas of potential hazard, but often quite difficult to predict when failure will occur and motion will begin. Obviously a trigger such as heavy precipitation or an earthquake can initiate movement, but even with the best modeling, surprises occur.

Three special topics boxes, describing effects of chemical fertilizers, tropical rainforests, and erosion in the Himalayas, stress the delicate relationships between human activities and soils.

Answers to Discussion Questions

1. (a) mechanical; (b) chemical (oxidation); (c) chemical (dissolution); (d) chemical (precipitation); (e) exfoliation is normally considered to be mechanical weathering, but if hydration is important as discussed in the text, chemical weathering also contributes to exfoliation; (f) again, the distinction between mechanical and chemical weathering is not always entirely clear-cut. Frost wedging is clearly mechanical, but it occurs when water freezes, which may be considered a chemical process.

2. In the Arctic, decay is so slow that little nutrient recycling and accumulation of humus occur. In the tropics, especially in tropical rainforests, decay, uptake of nutrients, and leaching are so rapid that soils are often poor because few nutrients remain in the soil. In the temperate regions, decay is relatively rapid in summer but slow in winter. This balance creates the richest soils. Some of the plant litter decomposes and nutrients are recycled, but at the same time a reservoir of litter and nutrients is retained in the soil.

3. Modern farming practices have been successful. Today, agricultural productivity is high in developed countries and food is inexpensive. Many people argue that chemical fertilizers are necessary to feed the present world population. The opposing argument is that chemical farming methods deplete soil and are not sustainable. Also chemical pollution of surface and groundwater results from chemical fertilizers (See Chapter 11). This argument continues that it is possible to maintain high yields economically without using chemical fertilizers. It is often cheaper to spread chemical fertilizers than to compost organic wastes and use them as fertilizer. However, if the cost of disposing these wastes is added to the cost of chemical agriculture, then in some instances, the balance shifts in favor of organic farming.

4. The decay of parent rock, accumulation of organic matter, erosion, and deposition are all time related factors as explained in the text.

5. Exfoliation plates may be further loosened by frost wedging, and then fall or slide under the influence of gravity.

6. Mass wasting is uncommon on the moon as evidenced by prominent 4 billion year old meteorite craters. As mentioned in the question, gravitation is weaker on the Moon than on the Earth. In addition, most mass wasting on Earth occurs after some other process destabilizes a slope. A stream or glacier may undermine a hillside, rainfall lubricates and adds weight to surface rock or regolith, earthquakes shake the surface, or volcanic eruptions melt large quantities of ice. These processes are nonexistent on the moon.

7. Earthflow, debris flow, and mudflow are common in deserts and in other dry areas with occasional heavy rainfall. In a typical scenario, the vegetation is destroyed by wildfires during the dry season. If heavy rain then falls on the barren ground, there are no roots to hold the soil and absorb moisture, so

mass wasting is likely. This sequence of events is common in the chaparral of southern California and other areas with alternate dry and wet seasons.

A similar scenario also occurs in wetter environments. A recent study shows that much of the surface topography in the Yellowstone Park region formed after catstrophic fires similar to the 1988 fire.

8. (a) No mass wasting; (b) Very susceptible if the layering dips parallel to the slope and if a stream flows at the base of the slope and undercuts it. Less susceptible if rocks dip away from the slope, or are horizontal; (c) again, very susceptible.

9. If this project is pursued, an additional exercise is to determine what types of building regulations (if any) apply to construction on potentially unstable slopes.

10. In Section 10.7 we cite the case of a mudflow that killed 20,000 people in Armero, Columbia when the volcano Nevado del Ruiz erupted nearly 50 kilometers away. The eruption itself occurred on an unpopulated mountain, and hence, did little or no human damage. But the mudflow followed stream valleys to populated areas. In 1970 an earthquake triggered a similar landslide from the 6663 meter-high top of Nevado Huascaran in central Peru. The flow buried the town of Yungay, 20 kilometers away, killing 17,000 people. In both cases, and in many similar ones, the earthquake or volcanic eruption caused little destruction or human death, but the resulting landslides were deadly.

Selected Bibliography

General texts on geomorphology are:

Dale Ritter: Process Geomorphology. Dubuque, IA, William C. Brown and Company, 1986. 592 pp.

Larry Mayer: Introduction to Quantitative Geomorphology. Englewood Cliffs, NJ, Prentice-Hall, 1989. 384 pp.

Michael A. Summerfield: Global Geomorphology: An Introduction to the Study of Landforms. New York, John Wiley and Sons, 1990. 537 pp.

Soil, soil erosion, and agriculture are discussed in:

Lester R. Brown and Edward C. Wolf: Soil Erosion: Quiet Crisis in the World Economy. Worldwatch Paper 60; Washington D.C., Worldwatch Institute, 1984. 49 pp.

Pierre R. Crosson and Anthony T. Stout: Productivity Effects of Cropland Erosion in the United States. Baltimore, MD, John Hopkins University Press, 1983. 103 pp.

Robert West Howard: The Vanishing Land. New York, Villard Books, 1985. 318 pp.

Sandra S. Batie and Robert G. Healy (eds.): The Future of American Agriculture as a Strategic Resource. Washington, D.C., The Conservation Foundation. 1980. 291 pp.

Wes Jackson, Wendell Berry, and Bruce Colman (eds.): Meeting the Expectations of the Land. San Francisco, North Point Press. 1984. 247 pp.

Edwin H. Clark II, Jennifer A. Haverkamp, and William Chapman: Eroding Soils, The Off-Farm Impacts. Washington, D.C., The Conservation Foundation. 1985. 252 pp.

Two useful, standard text on soils are:
Peter Birkelend: Soils and Geomorphology. London, Oxford University Press, 1984. 497 pp.

Ray Miller and Ray Donahue: Soils: An Introduction to Soils and Plant Growth. Englewood Cliffs, NJ, Prentice-Hall, 1990. 752 pp.

Recent articles on mass wasting:
Kenneth Hewitt: Catastrophic Landslide Deposits in the Karakoram Himalaya. Science, 242:64, 1988.

David K. Keefer, Raymond C. Wilson, Robert K. Mark, Earl E. Brabb, William M. Brown III, Stephen D. Ellen, Edwin, L. Harp, Gerald F. Wieczorek, Christopher S. Alger, and Robert S. Zatkin: Real-Time Landslide Warning During Heavy Rainfall. Science, 238:921, 1987.

Chapter 10

Weathering, Soil, and Erosion

Multiple Choice:

1. Weathering is the decomposition and disintegration of rocks and minerals at the Earth's surface by
(a) mechanical processes; (b) chemical processes; (c) both mechanical and chemical processes; (d) internal processes.

2. Agents of erosion include
(a) wind; (b) water; (c) glaciers; (d) gravity; (e) all of these.

3. Carvings on old headstones become faint and poorly preserved because of
(a) mechanical weathering; (b) chemical weathering; (c) deposition; (d) accumulation.

4. Atmospheric carbon dioxide dissolves in rainwater and reacts to form
(a) a weak acid; (b) a weak base; (c) a strong acid; (d) a strong base; (e) none of these.

5. When water and atmospheric gases attack granite, the feldspar crystals weather to form
(a) quartz; (b) sand; (c) gravel; (d) clay; (e) carbonic acid.

6. The least soluble component of granite is
(a) quartz; (b) feldspar; (c) calcite; (d) carbonate; (e) none of these.

7. ______ is a layer of loose rock fragments mixed with clay and sand overlying bedrock.
(a) Loam; (b) Humus; (c) Soil or regolith; (d) Litter; (e) Cap rock

8. Carbon, phosphorus, nitrogen, and potassium are examples of
(a) loam; (b) humus; (c) soil nutrients; (d) litter; (e) soil horizons.

9. If you dig down through a typical soil, you can see several layers called
(a) soil horizons; (b) soil nutrient; (c) regolith; (d) leaching zones.

10. The soil layer in which you would find the most organic material is the
(a) B horizon; (b) O horizon; (c) A horizon; (D) C horizon; (e) zone of leaching.

11. The horizon in which you would find the most clay and dissolved ions in a soil profile is the
(a) B horizon; (b) O horizon; (c) A horizon; (D) C horizon.

12. Tropical soils contain high proportions of
(a) iron; (b) aluminum; (c) iron and aluminum; (d) sodium and calcium; (e) soluble elements.

13. Soil on a hillside is generally _______ than on the valley floor.
(a) deeper and richer; (b) thinner and poorer; (c) younger and richer; (d) older and drier

14. Soil is usually only a ______ _____ thick or less in most parts of the world.
(a) hundred meters; (b) few meters; (c) few centimeters; (d) few kilometers

15. In nature, soil erodes
(a) more slowly than it forms; (b) approximately as rapidly as it forms; (c) faster than it forms; (d) in place and does not move.

16. The ______ is the maximum slope or steepness at which loose material remains stable.
(a) angle of repose; (b) mass wasting angle; (c) dip slope; (d) slip slope

17. During ______ loose, unconsolidated regolith moves downslope as a viscous fluid, analogous to road tar on a hot day or fluid magma pouring down the side of a volcano.
(a) slide; (b) fall; (c) flow; (d) slump

18. Typically, creep occurs at a rate of
(a) about 1 centimeter per year; (b) about 1 meter per year; (c) about 20 centimeters per year; (d) about 1 kilometer per year.

19. If earth material does not flow as a fluid, but rather slides downslope as a coherent mass, or as several blocks that remain intact, it is called
(a) slide; (b) slip; (c) fall; (d) flow; (e) solifluction.

20. A type of mass wasting that occurs when water-saturated soil moves over permafrost is called
(a) slide; (b) slip; (c) fall; (d) flow; (e) solifluction

21. The mass wasting that created devastation near the Madison River in Montana was triggered by
(a) improper irrigation; (b) an earthquake; (c) a volcano; (d) a careless motorist.

Chapter 10: Test

True or False:

1. Weathering involves little or no movement of the decomposed rocks and minerals.

2. Oxidation in rocks occurs when iron in minerals reacts with oxygen.

3. If you live in a moist climate, would be likely to find grains of halite in your backyard.

4. If water is acidic or basic, the resulting solution has a greater ability to dissolve rocks and minerals.

5. When acidic rainwater seeps into cracks in limestone, it dissolves the rock, enlarging the cracks to form subterranean caves and caverns.

6. The alternate shrinking and swelling of humus keeps soil loose, allowing roots to grow easily.

7. Nutrient abundance or deficiency does not depend on the chemical composition of the parent rock.

8. The most fertile soils are those of prairies and forests in temperate latitudes.

9. Soil continues to accumulate and thicken throughout geologic time.

10. In the United States, approximately one third of the topsoil that existed when the first European settlers arrived has been lost.

11. Orientation or layering in rock has no effect on slope stability.

12. Vegetation decreases slope stability.

13. Volcanic activity may initiate slides by melting snow and ice near the tops of volcanoes.

14. Fall is the slowest type of mass wasting.

15. Debris flows are common on volcanic slopes.

16. You can distinguish slump from creep by the orientations of the trees.

17. Added water seldom triggers landslides.

Completion:

1. Decomposition and disintegration of rocks and minerals at the surface by mechanical and chemical processes is _____.

2. The removal of weathered material is _____.

3. The physical disintegration of rocks is _____ _____.

4. Alteration of the composition of rocks and minerals by the interaction of air and water at the Earth's surface is _____ _____.

5. Rusting is a form of _____ weathering.

6. As feldspar weathers to clay, the clay crystals take on water in a process called ______.

7. The most fertile soils, called ______ are composed of a mixture of sand, clay, silt, and generous amounts of organic matter.

8. The B soil horizon is also called the zone of _____.

9. ______ is a hard cement that forms when calcium carbonate precipitates in the soil.

10. A highly aluminous soil, called ______ is the world's main source of aluminum ore.

11. Movement of material downslope primarily under the influence of gravity is called _____ _____.

12. Slow downhill movement of rock or soil is _____.

13. A flowing slurry of clay, silt, sand and rock is a/an _____ _____.

14. Water-saturated soil moves over frozen ground water by _____.

15. ______ is the rapid movement of a newly detached segment of bedrock.

Answers for Chapter 10

Multiple Choice: 1. c; 2. e; 3. b; 4. a; 5. d; 6. a; 7. c; 8. c; 9. a; 10. b; 11. a; 12. c; 13. b; 14. b; 15. b; 16. a; 17. c; 18. a; 19. a; 20. e; 21. b; 22. b

True or False: 1. T; 2. T; 3. F; 4. T; 5. T; 6. T; 7. F; 8. T; 9. F; 10. T; 11. F; 12. F; 13. T; 14. F; 15. T; 16. T; 17. F

Completion: 1. weathering; 2. erosion; 3. mechanical weathering; 4. chemical weathering; 5. chemical; 6. hydration; 7. loams; 8. accumulation; 9. caliche; 10. bauxite; 11. mass wasting; 12. creep; 13. debris flow; 14. solufluction; 15. rockslide

CHAPTER 11

Streams and Ground Water

Discussion

Many geological processes occur simultaneously. However, a professor or a book can only teach one concept at a time. In many cases, problems arising from this dilemma are easily resolved and a logical pedagogic order reveals itself. However, in discussing stream behavior it seems as though everything must be taught first. A stream doesn't first erode sediment, then transport it, and finally deposit it. Rather all three processes may occur at the same time in almost the same location. Similarly changes in channel shape, sediment load, discharge, velocity, and tectonic rejuvenation are so closely interrelated that it becomes misleading to separate the topics into discrete sections. There is no choice but to teach one topic at a time, but it is also important to remind the student that nature is not always as sequential as a textbook.

The term 100-year flood is commonly used, and refers to a flood level that has a 1 percent chance of occurring in any given year. If a 100 year flood occurs in 1990, the probability of recurrence in 1991 is 1/100, just as in any other year. The mathematics is just the same as in tossing a coin; the probability of having the coin land heads up is 50 percent, no matter what preceded any given toss.

The instructor might introduce placer deposits from Chapter 22, in the discussion of deposition in stream beds.

Significant legal and economic issues are involved in the diversion of water from one source to another. In a landmark case, The Los Angeles Department of Water and Power (LADWP) removed so much water from streams on the eastern slopes of the Sierras that some dried up completely. One of these streams was Rush Creek, which fed Mono Lake, a playa lake with no outlet. Deprived of most of its natural inflow, the lake level dropped by 12 meters, its salinity increased, and many plants and animals died. The National Audubon Society brought suit to halt the removal of water from the creek.

The Audubon Society claimed that the public trust takes precedence over all rights to appropriate water. The LADWP argued that the regulations permitting the withdrawal of water had replaced the public trust doctrine and that the water could be withdrawn indefinitely without concern for the environmental consequences. The court said, "We are unable to accept either position." Therefore, an accommodation should be reached between the two viewpoints. If the Audobon Society's position on withdrawal of water were broadly and totally upheld, the harm to the city of Los Angeles would be unacceptable. However, the State has a duty to "protect public trust uses wherever feasible." The Court recognized that no such consideration to environmental consequences had ever been given. Therefore, according to the Court, "Some responsible body ought to reconsider the allocation of the waters of Mono basin." This decision was not a clear win for either side and the dispute has not yet been completely resolved.

In recent years, pollution of ground water has become an important public issue. As a result, environmental ground water hydrology has become an increasingly important career opportunity for geologists. Therefore, after introducing the basic concepts, we have emphasized the environmental theme in this chapter.

Section 11.11, describing the Love Canal disaster is included to show the integration of science and public policy. Special topics boxes on depletion of the High Plains (Ogallala) aquifer, and on soil salinization resulting from irrigation, stress the effects of excessive water exploitation.

As an additional exercise, the instructor may encourage students to research local ground water quality. The Environmental Protection Agency or other government agencies may have analyzed well water for biological and chemical pollutants. One student may be assigned to interview the representatives of the EPA. Some questions might include: How many hazardous waste disposal sites exist in the area and how do they effect ground water? How many of the local gas stations have leak detectors on their tanks and how old are the tanks? How is municipal garbage disposed of; if there is a public landfill, do proper precautions exist to protect ground water?

Answers to Discussion Questions

1. (a) If global temperature rose, more water would evaporate from the oceans and therefore precipitation and runoff on the continents would increase. Climatologists also calculate that global warming might lead to changes in wind and current patterns. Monsoon cycles in Asia and jet streams in higher latitudes might be altered. If so, rainfall patterns would change, humid areas might be desertified, deserts could receive more rainfall. These changes could affect agriculture and the economy in many areas. A rise in temperature might also lead to melting of the Greenland and Antarctic ice caps and a rise in sea level, although this scenario is uncertain. The effect of melting of coastal glaciers could be offset if increased precipitation led to an increase in the amount of permanent ice on continents. The situation is further complicated by the fact that the climate in some areas in Northern Canada and the USSR is cold enough for glaciation if more moisture were available. Therefore, paradoxically, a warming trend could initiate glaciation in some places.

(b) If global temperature fell, less water would evaporate from the oceans and therefore precipitation and runoff on the continents would decrease. Semiarid lands could be desertified. Again patterns of precipitation might change, with unknown consequences. Ice would be retained longer and in some areas glaciers would expand.

2. If the stream is constricted, the velocity at the constriction would increase. This increase in velocity would increase downstream erosion. The channel would deepen or widen, or both, downstream. Upstream, sediment might accumulate as the current slowed.

3. Because the Selway is a small river in the mountains, it responds rapidly to rising temperature in the spring (which melts snow) and to rainfall. As

explained in the text, the flow of desert streams may fluctuate even more dramatically, whereas streams farther from the mountains or those that draw their water from a larger drainage basin often fluctuate less.

4. Erosion increases during flood because the velocity and discharge both increase dramatically. Therefore, during flood both the competence and capacity of a stream reach their maximums.

5. A natural levee forms because sediment carried by a stream is deposited where the stream velocity decreases as the water overflows its banks. Thus, during the formation of natural levees, some sediment is removed from the channel. However, as explained in the text, artificial levees cause sediment to accumulate in a stream channel. Over the short term, artificial levees prevent flooding. However, over many years, the increased sedimentation in the channel can raise the channel above the floodplain, creating the danger of truly cataclysmic floods. The paradox is that if a stream is allowed to flood periodically, it will cause less damage in the long-term. But people who formulate public policy are often under pressure to seek short-term solutions and must address the question, "Can you do something to save my home or business now or next year, not 5, 25, or 100 years from now?"

6. (a) Dendritic; (b) trellis; (c) radial.

7. Both wells in the first drawing could have water. Well (b) could be polluted because waste from the septic tank could flow across the surface of the impermeable shale to the well. Well (a) is artesian and is isolated from the septic tank by the shale. However, we don't have any information about other potential sources of pollution. In the second drawing, well (c) could be polluted because contaminated water could flow rapidly through the interconnected fractures in the granite. Well (d), in an unfractured portion of granite, would not produce water.

8. This type of failure is common and could occur if the well were dug into impermeable rock or regolith such as unfractured granite, clay, or shale.

9. Streams in humid areas lie at or below the water table, and consequently receive water from ground water during dry seasons. In contrast, many desert streams lie above the water table and lose water into the ground water reservoir.

10. In support of Reisner, one can argue that water is being removed from western aquifers faster than the rate of recharge, so it is inevitable that ground water resources will be depleted. Since not enough water flows in surface streams to maintain existing levels of irrigation, failure of the agricultural system is inevitable. Furthermore, soil salinity increases when fields are irrigated year after year and this problem is not easily reversed. The argument against Reisner is that we have technology not available to the ancients. Drainage systems can be built and salt-resistant crops can be genetically engineered. Water could be diverted from British Columbia, where there is an excess. However, large amounts of energy are needed for such

projects. Other technological solutions would be to develop and use drought-resistant strains of grain, or to use water more efficiently with advanced irrigation techniques.

11. Neither granite nor shale are readily soluble so caverns are much less common in these rocks. However, caves form occasionally from weathering of fractures or joints, and from collapse of large blocks of rock.
12. If calcium carbonate precipitates from a pool or stream it will settle out in a broad sheet on the floor of a cavern. Stalactites and stalagmites form as water drips from the ceiling of a cavern. As water droplets form and fall, the calcium carbonate precipitates in icicle-like structures.

13. Ground water moves relatively slowly through a sandstone aquifer, allowing time for natural processes to degrade pollutants. In karst topography, underground streams flow rapidly. More oxygen is available and rates of mixing are high, but the flow is so fast that pollutants are carried great distances before they break down.

Selected Bibliography

Refer to the geomorphology texts cited in Chapter 10.

Water use and diversion projects:
Helen Ingram: Water Politics: Continuity and Change. Tucson, University of Arizona, 1990. 158 pp.

Marc Reisner and Sarah Bates: Overtapped Oasis: Reform or Revolution for Western Water. New York, Island Press, 1990. 200 pp.

David A. Franko and Robert G. Welch: To Quench Our Thirst, The Present and Future Status of Freshwater Resources of the United States. Ann Arbor, University of Michigan Press, 1983. 148 pp.

Marc Reisner: Cadillac Desert. New York, Viking Press, 1986. 582 pp.

Kai N. Lee: The Columbia River Basin. Environment, 31:6, (7), 1989.

Taylor O. Miller, Gary D. Weatherford, and John E Thorson: The Salty Colorado. Washington, D. C., The Conservation Foundation, 1986. 102 pp.

A survey of water quality:
Richard A. Smith, Richard B. Alexander, and M. Gordon Wolman: Water-Quality Trends in the Nation's Rivers. Science, 235:(1607), 1987.

General books on ground water and karst topography:
Roscoe Moss Company: Handbook of Ground Water Development. New York, John Wiley and Sons, 1989. 900 pp.

William B. White and Elizabeth L. White, eds.: Karst Hydrology: Concepts from the Mammoth Cave Area. New York, Van Nostrand Reinhold, 1989. 346 pp.

Pollution of groundwater and its consequences:

Robert M. Clark, Carol Ann Fronk, and Benjamin W. Lykins, Jr.: Removing Organic Contaminants From Groundwater. Environmental Science and Technology, 22:10, (1126), 1988.

Penny Newman: Cancer Clusters Among Children: the Implications of McFarland. Journal of Pesticide Reform, 9:3, (10), 1988.

Ruth Patrick, Emily Ford, and John Quarles: Groundwater Contamination in the United States, 2nd edition. Philadelphia, PA, University of Pennsylvania Press, 1987. 647 pp.

Geothermal energy:

R. Monastersky: Drilling Begins in Search of Molten Energy. Science News, 136:(101), 1989.

Chapter 11

Streams and Ground Water

Multiple Choice:

1. Precipitation includes
(a) rain; (b) snow; (c) hail; (d) sleet; (e) all of these.

2. Water that flows back from the land to the sea is
(a) oceanic; (b) runoff; (c) transpiration; (d) precipitation; (e) evaporation.

3. What determines the velocity of a stream?
(a) gradient of the stream bed; (b) flow o discharge the stream; (c) shape of the channel; (d) all of these

4. The volume of water flowing downstream per unit time is
(a) precipitation; (b) discharge; (c) gradient; (d) flood; (e) tributary.

5. The ability of flowing water to dislodge fragments of rock and grains of sediment is called
(a) abrasion; (b) solution; (c) hydraulic action; (d) competence; (e) none of these.

6. The total amount of sediment a stream can carry past a point in a given amount of time is called
(a) abrasion; (b) capacity; (c) hydraulic action; (d) competence; (e) none of these.

7. The ions carried in solution in a stream constitute the
(a) abrasion load; (b) suspended load; (c) bed load; (d) dissolved load; (e) saltation factor.

8. How is most material carried in a stream?
(a) saltation; (b) traction; (c) solution; (d) suspension; (e) abrasion

9. The water on the outside of a stream curve moves
(a) faster than the water on the inside; (b) slower than the water on the inside; (c) the same speed as the water on the inside; (d) mostly vertically in a process called upwelling.

10. When a stream has a greater supply of sediment than it can carry,
(a) it erodes some of the sediment in its channel; (b) it deposits some of the sediment in its channel; (c) it deposits some of the sediment on its banks; (d) it stops flowing.

11. The portion of a valley covered by water during a flood is called the
(a) natural levee; (b) alluvial plain; (c) delta; (d) flood plain; (e) meander.

12. The ultimate base level of a stream is
(a) the low-water benchmark; (b) flood stage; (c) natural levee; (d) sea level; (e) elevation of the confluence.

13. The top of a waterfall is an example of
(a) ultimate base level; (b) temporary or local base level; (c) the low-water benchmark; (d) a natural levee; (e) none of these.

14. Low-gradient streams are likely to carve
(a) V shaped valleys; (b) U shaped valleys; (c) numerous waterfalls; (d) valleys with meanders and oxbow lakes in their flood plains; (e) steep gullies.

15. The region ultimately drained by a single river is a
(a) flood plain; (b) channel; (c) drainage divide; (d) drainage basin; (e) base level.

16. When headward erosion cuts through a drainage divide into the basin on the other side of the divide, the stream
(a) eventually captures the drainage of the other basin; (b) reverses direction and flows the other way; (c) stops cutting backward; (d) none of these.

17. Sedimentary rocks are
(a) generally less porous than igneous rocks; (b) generally more porous than igneous rocks; (c) equally porous as igneous rocks; (d) less likely to contain ground water than igneous rocks.

18. The region of the crust in which all pore spaces contain water is
(a) asthenosphere; (b) zone of accumulation; (c) zone of ablation; (d) zone of saturation; (e) zone of aeration.

19. In a typical aquifer, ground water flows at
(a) the rate of surface rivers; (b) about 4 m per day; (c) about 15 m per year; (d) 1 to 2 cm per year; (e) twice the rate of surface runoff.

20. Sinking or settling of the Earth's surface
(a) can be caused by removal of ground water; (b) is subsidence; (c) results from collapse of the pore spaces in an aquifer; (d) can not usually be reversed; (e) all of these.

21. Which is biodegradable in ground water?
(a) normal sewage; (b) DDT, a pesticide; (c) paint; (d) solvents;

(e) arsenic.

22. Caverns usually form in
(a) granite; (b) limestone; (c) fractured schist; (d) conglomerate; (e) shale.

23. Karst topography
(a) forms in high granite mountains; (b) is characteristic of limestone regions with abundant caves; (c) is a featureless coastal plain; (d) is most common in sandstone; (e) forms best in shales.

24. Which of the following statements about geothermal energy is true?
(a) Geothermal energy contributes a large portion of U.S. energy demand. (b) Geothermal energy has been proved in experimental sites but has not been developed commercially. (c) Commercial geothermal energy production involves drilling wells into hot, dry rock. (d) Commercial geothermal energy production involves drilling wells into areas with hot ground water reserves.

True or False:

1. The velocity of a stream increases when discharge increases.

2. Water flows more rapidly near the banks than near the center of a stream.

3. Clay and silt are small enough that even the slight turbulence of a slow stream keeps them in suspension.

4. When stream current slows down, the stream loses its ability to transport the largest particles and deposits them in the stream bed.

5. A stream erodes its bank on the inside of a curve.

6. Braided streams are rare in both deserts and glacial environments because they both lack sediment.

7. A 100-year flood happens regularly every 100 years.

8. An idealized graded stream is one in which gradient, discharge, and current velocity are in equilibrium.

9. Streams will not commonly follow faults.

10. Globally, 30 times more water is stored as ground water than in all streams and lakes combined.

11. If you dig into the unsaturated zone, the hole will fill with water.

12. The water table rises and falls with the seasons.

13. Subsidence is a readily reversible process.

14. Polluted ground water may be purified slowly by natural processes.

15. The problems that took place at Love Canal in New York are an example of ground water pollution.

16. Sinkhole formation cannot be intensified by human activities.

17. The water in a geyser is hot but not quite boiling.

18. The United States is the smallest producer of geothermal electricity in the world.

Completion:

1. During a flood, a stream overflows its banks and water covers the adjacent ______ ______.

2. The ______ of a stream is a measure of the largest particle it can carry.

3. ______ is a mode of transport in which water turbulence keeps fine particles mixed with the water and prevents them from settling to the bottom.

4. If the stream velocity is sufficient, sand grains bounce along in a series of short leaps or hops called ______.

5. A/an ______ is an elongate mound of sediment in a stream channel.

6. ______ ______ are fan-shaped deposits of multi-sized particles.

7. Downward stream erosion is called ______.

8. In ______ patterns stream tributaries resemble veins in a leaf.

9. A stream is ______ when any change increases its erosive ability.

10. A/an ______ stream is established before local uplift starts and cuts its channel at the same rate that the land is rising.

11. ______ is the proportion of a volume of rock or soil that consists of open spaces.

12. The ______ ______ is the top of the zone of saturation.

13. A/an ______ is any body of rock or regolith that can yield economically significant quantities of water.

14. If water is withdrawn faster than it can flow into a well a/an ______ ______ ______ forms.

15. Mineral deposits formed in caves by the action of water are called ______.

16. A/an ______ forms when the roof of a limestone cavern collapses.

17. A/an ______ erupts hot water and steam violently onto the Earth's surface.

18. Hot springs have been tapped to produce ______ ______.

Answers for Chapter 11

Multiple Choice: 1. e; 2. b; 3. d; 4. b; 5. c; 6. b; 7. d; 8. d; 9. a; 10. b; 11. d; 12. d; 13. b; 14. d; 15. d; 16. a; 17. b; 18. d; 19. c; 20. e; 21. a; 22. b; 23. b; 24. d

True or False: 1. T; 2. F; 3. T; 4. T; 5. F; 6. F; 7. F; 8. T; 9. F; 10. T; 11. F; 12. T; 13. F; 14. T; 15. T; 16. F; 17. F; 18. F

Completion: 1. flood plain; 2. competence; 3. suspension; 4. saltation; 5. bar; 6. alluvial fans; 7. downcutting; 8. trellis; 9. rejuvenated; 10. antecedent; 11. porosity; 12. water table; 13. aquifer; 14. cone of depression; 15. speleothems; 16. sinkhole; 17. geyser; 18. geothermal energy

CHAPTER 12

Glaciers and Wind

Discussion

We complete our discussion of surface processes with a chapter on glaciers and wind. While a glacier is a mass of solid ice, capable of eroding and transporting massive boulders, only the smallest and lightest particles can be lifted and transported very far by wind. Yet the two are related because both are significant geologic agents in special climatic environments with only sparse vegetation.

In discussing the movement of glaciers we utilize two concepts from our previous discussions of rocks and plate tectonics: phase changes initiated by changes in pressure, and plastic behavior of crystalline solids.

In Chapter 7, we showed that hot mantle rock melts when pressure is reduced. In this chapter we explain that an increase in pressure melts ice. There is no contradiction here, the difference arises because rock expands as it melts, whereas ice contracts as it melts. As a demonstration, suspend an ice cube in a freezer and hang a thin wire over it. Then weight the wire on both ends and leave it for a week. The pressure melts the ice directly under the wire. The meltwater refreezes above the wire as soon as the pressure is removed. Eventually the wire migrates entirely through the ice cube but does not cut it in half.

It is more difficult to illustrate the plastic behavior of snow and ice, especially for students not accustomed to cold winters. However, if you live in an area with abundant snowfall, you need only tell students to observe snow curling as it slowly creeps off a roof to convince them of the plasticity of ice.

This is a good time to reinforce the concept that the two million year time span of Pleistocene glaciation is a thin slice of geologic time. The first tool-making human ancestors evolved roughly two million years ago. Recall, that if geological time were measured as the distance from an English king's nose to the end of his outstretched fingers, two million years can be removed by one swipe of a nail file.

In the box Earth Science and the Environment, The Dust Bowl, we discuss problems that can result from agriculture in semiarid regions. In many such regions, human pressure is manifested in other ways than agriculture. Over long weekends in the fall and winter, as many as 70,000 people may visit a narrow strip of sand dunes on the eastern edge of the Imperial Valley in California. Many ride motorcycles or dune buggies. The tire tracks disrupt vegetation and leave deep trenches in nearby playas. Chapter 12 also contains an additional Earth Science and the Environment box on surging glaciers.

Answers to Discussion Questions

1. (a) The glacier would retreat; (b) The glacier would advance; (c) The glacier would advance, especially if the summer temperature decreased.

2. Not enough snow falls in winter, so all the snow melts even though the summers are relatively short and cool.

3. The lower flanks of the mountains may have been recently scoured by alpine glaciers.

4. Compare the rock type of the boulder with that of the country rock, look for other evidence of glaciation, such glacial till or striations on the boulders or on bedrock. Search in the direction that a glacier came from for country rock similar to that of the boulder.

5. Moraines are typically unsorted and unstratified, stream sediment is typically sorted and layered. Sometimes the spaces between stream cobbles fill with sediment. However, if this occurs, the cobbles will still be in contact with one another. In a ground moraine, cobbles are commonly do not touch, but are suspended in a finer matrix. Angular cobbles and boulders are often deposited by a glacier. However, not all rounded rocks are deposited by streams: often glaciers transport and deposit sediment that was previously rounded during stream transport. Look for morainal landforms.

6. Medial moraines form when glaciers merge, and they can merge only if they are moving.

7.

	Erosion	Transport	Deposition
Wind	Lifts only small particles. Wind often erodes large areas of unvegetated land.	Silt can be carried aloft but sand is carried mostly by saltation. Grains larger than sand are not transported.	Typical wind deposited sediment is fine-grained loess or migrating dunes.

Streams	More viscous than wind and can erode even large cobbles and boulders during flood. Most erosion occurs in channel.	Because gradient is generally steeper in mountainous areas, large particles are transported near head-waters and only small ones in valleys.	Sediment is sorted as large particles are deposited near headwaters and smaller ones transported farther toward oceans.
Mass Wasting	Erosion occurs mainly on steep hill-sides. All particle sizes are eroded together.	All sizes transported down steep hillsides.	Many landslides deposit debris in an unsorted jumble, al-though earth-flow and mud-flow transport mainly smaller grains.
Glaciers	Erodes all sizes from clay to boulders. Occurs only in cold wet environments.	Ice is viscous enough to transport all particle sizes.	Deposits unsorted sediment when ice melts.

8. Alluvial fans typically contain cobbles that are too large to be transported by the wind.

9. (a) parabolic; (b) barchan.

10. You can deduce that there was a plentiful supply of sand or silt and wind to transport it. If there is wind, an atmosphere exists. Other conclusions would be tentative. For example it would not necessarily be correct to conclude that the dunes formed in a desert. On Earth, dunes can also form along coastlines where vegetation is sparse.

Chapter 12

Selected Bibliography

Refer to the geomorphology texts cited in Chapter 10.

A general text on glaciers:
Robert P. Sharp: Living Ice. Cambridge, Cambridge University Press, 1988. 225 pp.

A detailed study of glacial movement is given in:
Hermann Engelhardt, Neil Humphrey, Barclay Kamb, and Mark Fahnestock: Physical Conditions at the Base of a Fast Moving Antarctic Ice Stream. Science, 248:57, 1990

Coastal dunes are discussed in:
Karl Nordstrom, et al: Coastal Dunes, Form and Process. Somerset, N.J., John Wiley and Sons, 1990. 392 pp.

Chapter 12

Glaciers and Wind

Multiple Choice:

1. Glaciers form in regions with
(a) cold and wet climates; (b) cold and dry climates; (c) warm and wet climates; (d) warm and dry climates; (e) all of these.

2. The ice sheets of Greenland and Antarctica contain ___ % of the world's ice.
(a) 100; (b) 99; (c) 50; (d) 25; (e) 5

3. In some instances, a glacier may surge at speed of
(a) 10 to 50 centimeters per day; (b) 10 to 50 kilometers per day; (c) 10 to 50 meters per day; (d) 10 to 50 meters per year.

4. Near the surface of a glacier the ice is
(a) brittle; (b) liquid; (c) always covered by snow; (d) under high pressure; (e) a and c.

5. The zone above the firn line is
(a) the zone of accumulation; (b) the ablation area; (c) the zone of aeration; (d) snow line; (e) the zone of wastage.

6. The down-valley end of a glacier is the
(a) bergshrund; (b) lateral moraine; (c) firn line; (d) wastage line; (e) terminus.

7. Grooves and scratches made by a glacier are
(a) glacial polish; (b) plucking; (c) rock flour; (d) glacial striations; (e) roche moutonnee.

8. A narrow ridge between glacial valleys is a/an
(a) bergshrund; (b) cirque; (c) horn; (d) arete; (e) roche moutonnee.

9. A boulder that differs from the bedrock in the immediate vicinity is a/an
(a) erratic; (b) roche moutonnee; (c) tarn; (d) till; (e) drumlin.

10. When a glacier is at its greatest advance, it deposits a (a) medial moraine; (b) terminal moraine; (c) recessional moraine; (d) ground moraine; (e) lateral moraine.

11. A snake-like ridge that forms as the bed deposit of a stream that flowed within or beneath a glacier is
(a) a kame; (b) an esker; (c) a valley train; (d) an outwash

plain; (e) a drumlin.

12. The most recent ice age was
(a) in the Cambrian; (b) in the Pleistocene; (c) in the late Paleozoic; (d) 250 years ago; (e) 2.3 billion years ago.

13. When glaciers grow sea level
(a) falls; (b) rises; (c) remains constant; (d) rebounds; (e) none of these.

14. Erosion of soil by wind is
(a) saltation; (b) deflation; (c) inflation; (d) minimal in deserts; (e) augmented by plant cover.

15. Wind-scoured depressions are called
(a) dust bowls; (b) blowouts; (c) kettles; (d) playas; (e) sinkholes.

16. Wind carries sand by
(a) suspension; (b) saltation; (c) gliding; (d) solution; (e) slipping.

17. A crescent-shaped sand dune that forms in area of little sand is a _____ dune
(a) parabolic; (b) transverse; (c) barcan; (d) longitudinal; (e) stable

18. Wind blows sand out of an unvegetated depression and deposits it forms a
(a) parabolic dune; (b) transverse dune; (c) barcan dune; (d) longitudinal dune; (e) stable dune.

19. Loess is
(a) porous; (b) wind blown silt; (c) unlayered; (d) likely to form vertical cliffs and bluffs; (e) all of these

20. The largest loess deposits in the world are found in central
(a) New York; (b) China; (c) deserts; (d) Peru; (e) Takla Makan

True or False:

1. A glacier forms wherever the amount of snow that falls in winter exceeds the amount that melts in summer

2. The Greenland sheet is more than 2.7 kilometers thick in places and covers 1.8 million square kilometers.

3. Rates of glacial movement do not depend on steepness, precipitation, or temperature.

4. Pressure near the base of a glacier causes ice at the base of a glacier to melt.

5. Ice near the base of a glacier fractures to form crevasses.

6. Glaciers grow and shrink.

7. A flowing glacier is too slippery to scour areas of bedrock and erode landscapes.

8. Frost wedging releases rocks from the cirque walls.

9. Glacier forms V-shaped valleys.

10. Large glacial deposits found in some places were carried by icebergs during catastrophic floods.

11. Ice is so much more viscous than water that it carries particles of all sizes together.

12. The extent of continental glaciers of the most recent ice age can be determined by locating their terminal moraines.

13. Geologic evidence shows that the Earth has been cold and ice covered for about 90 percent of the past 2.5 billion years.

14. Because air is much less dense than water, wind can move only small particles, mainly silt and sand.

15. Windblown sand and silt are not effective agents of erosion.

16. Migrating dunes can overrun buildings and highways.

17. When a barchan dune migrates, the edges move slower because there is more sand to transport.

18. Much of the rich soil of the central plains of the United States formed on loess.

Completion:

1. A/an ______ is a massive, long-lasting accumulation of compacted snow and ice that forms on land.

2. A glacier that covers an area of 50,000 square kilometers or more is a/an ______ ______.

3. The entire glacier slides over bedrock by ______ ______ like a bar of soap on a tilted board.

4. ______ form in the upper 50 meters of a glacier as it flows over bedrock.

5. Giant chunks of ice ____ off glaciers to form icebergs.

6. Glaciers pry bedrock fragments loose and incorporates them in the ice in a process called ______.

7. If silt or sand, rather than rock, is embedded in the base of a glacier, it abrades a smooth, shiny finish called ______ ______.

8. A/an ______ is a steep-walled semicircular depression eroded into mountain by a glacier.

9. A small glacial valley lying high above the floor of the main valley is called a/an ______ ______.

10. ______ ______ is sediment that was first carried by a glacier and then transported and deposited by a stream.

11. A/an ______ is a mound or ridge of till deposited by a glacier.

12. Glacial streams deposit their sediment downstream from the glacier as ______.

13. A time of extensive glacial growth, when alpine glaciers descend into lowland valleys and continental glaciers spread over higher latitudes, is called a/an ______ ______.

14. When wind blows, it removes only the small particles, leaving the pebbles and rocks to form a continuous cover of stones called ______ ______.

15. The leeward face of a dune is called the ______ ______.

16. ______ ______ are common in moist semidesert regions and along seacoasts.

Answers for Chapter 12

Multiple Choice: 1. a; 2. b; 3. c; 4. e; 5. a; 6. e; 7. d; 8. d; 9. a; 10. b; 11. b; 12. b; 13. a; 14. b; 15. b; 16. b; 17. c; 18. a; 19. e; 20. b

True or False: 1. T; 2. T; 3. F; 4. T; 5. F; 6. T; 7. F; 8. T; 9. F; 10. F; 11. T; 12. T; 13. F; 14. T; 15. F; 16. T; 17. F; 18. T

Completion: 1. glacier; 2. ice sheet or continental glacier; 3. basal slip; 4. crevasses; 5. calve; 6. plucking; 7. glacial polish; 8. cirque; 9. hanging valley; 10. stratified drift; 11. moraine; 12. outwash; 13. ice age; 14. desert pavement; 15. slip face; 16. parabolic dunes

CHAPTER 13

Oceans and Coastlines

Discussion

This chapter introduces many elements of oceanography. It begins with the question, "Why does the Earth have oceans?", and then discusses seawater, waves, tides, and currents. While the similarities and differences between ocean currents and stream currents are obvious, an interesting contrast exists between ocean waves and stream waves. If a stream flows swiftly over a rock, a wave forms over the rock. In this case, the wave doesn't move, but the water flows through it and continues downstream. In contrast, an ocean wave moves across the surface of the sea, but the water doesn't move along with the wave.

There is considerable folklore concerning the Coriolis effect. One common belief is that pigs' tails curve clockwise in the Northern Hemisphere and counterclockwise south of the Equator. This is not supported by observation. A second popular belief is that the water draining out of a bathtub swirls clockwise in the Northern Hemisphere and opposite in the south. A 1962 article in Nature showed that in a normal bathtub residual currents from the original filling, air currents in the room, temperature differences, or the shape of the tub overwhelmed the Coriolis force. However under carefully controlled conditions, water swirled clockwise out of a symmetrical basin in Cambridge Mass.

The second half of the chapter is devoted to coastlines. It starts with the major features of coastlines, and then discusses coastline erosion and human settlement. Coastal erosion has become an increasingly important problem in recent years. In June, 1990, an article in Geotimes reported that rates of coastal erosion in North America range from negligible in many regions to losses of more than 10 meters of shoreline per year along sections of barrier islands in Virginia and Louisiana. In most places, erosion is greatest in undeveloped areas, but in Ocean City, Md. and Miami Beach, Fla., buildings constructed 30 to 50 years ago in seemingly safe regions are now threatened. A summary of erosion and accretion on various North American coastlines is given on the next page:

Region	Coastline with measurable erosion	Coastline with measurable accretion	No data
East coast	56%	19%	24%
West coast	63%	13%	24%
Gulf coast	55%	16%	28%
Great Lakes	47%	0%	52%

Chapter 13 contains two Earth Science and the Environment boxes: Energy from the Ocean and Pollution of the Oceans.

Answers to Discussion Questions

1. The properties of waves are discussed in Chapter 6 under Earthquakes and again here in Chapter 20 under Coastlines. The repetition is intentional. All waves reflect, refract, and interfere (interference is not discussed in the text). Thus, earthquake waves in rock, ocean waves in water, and electromagnetic waves in a vacuum behave in much the same manner. However all waves are not the same. A P wave is a compressional body wave in solid material or liquid. An S wave is a shear wave in a solid. An L wave is a surface wave, similar, but not identical to a water wave. Students find it confusing to say that different types of waves are different from one another and they affect the media in which they travel differently, yet they all reflect, refract, and interfere. Often they seek a simple answer, "Are different types of waves the same or are they different." The only accurate answer is: "Yes, they are the same, but they are also different."

2. The ship rides on the surface of the wave whereas waves break on the beach. If winds are strong enough, waves will break on the open ocean and these breaking waves cause the most damage to ships at sea.

3. On a shallow coastline, large waves encounter the sea floor and break before they reach shore.

4. The bottom of a wave encounters the sea floor when the depth diminishes to about one half the wavelength. The wavelength can be measured from aerial photos and the water depth can therefore be determined by observating the surf.

5. Spilled petroleum can be contained and recovered most easily if it remains concentrated. Mid-ocean currents have little or no affect along shorelines, where most tanker accidents occur. Longshore currents, tides, and storm waves all disperse spilled oil. As a result it is important to act as quickly as possible

when an accident occurs. When the Exxon Valdez struck a rock near Valdez, Alaska, crews were unable to deploy booms for several days. Critics argue that this delay greatly magnified the effects of the spill. One solution would be to require all tankers to carry booms and other containment devices on board. These could then be deployed immediately with the ship's lifeboats.

6. Both types of erosion are caused by moving water and by water-transported sediment. Both can alter landforms by erosion, transport, and deposition, and both undercut bedrock and cause mass wasting. Ocean waves, especially storm waves, have more energy than most stream currents. In addition, ocean waves affect an entire coastline and whereas stream erosion is confined to its channel or floodplain.

7. Tectonic activity and sea level changes constantly alter coastlines. The same question could also apply to mountain ranges. We could ask, "If mountains are being eroded, why do they still exist?"

8. Submergent coastline. Sea level has been rising since the last of the Pleistocene ice sheets began to melt. Submergence from rising sea level may be augmented by tectonic sinking.

9. The government should support the construction of groins: The government has an interest in protecting commerce and the lives of its citizens. Coastal stabilization, flood control projects, soil conservation, and avalanche control along highways all fall under this category.

The government should outlaw groins: In the long run, groins cause more erosion than they prevent. Furthermore, if one person builds a groin, he or she is effectively stealing the neighbor's beach and the government has a right to prevent one citizen from harming another.

The government should permit people to construct groins on their own property: The government should not interfere with peoples' lives or tell them want they can or cannot build on their own property.

10. As mentioned above, the government has an obvious interest in protecting commerce and the lives of its citizens. Various types of disaster relief are a form of national insurance. It is virtuous to help people who suffer losses from hurricanes, floods, drought, earthquakes, volcanic eruptions, or other natural disasters. Furthermore, disaster relief speeds the return to normal commerce and therefore helps the economy. An opposing argument is that it is common knowledge that barrier islands are geologically unstable, are prone to erosion, and are vulnerable to storms. In most areas, only wealthy people can

afford waterfront property. Why should the government protect rich people who court disaster by building in especially vulnerable regions?

11. A social solution would be to eliminate all forms of beach stabilization and then either accept periodic structure damage during large storms or abandon structures on the beach altogether. A national coastline park and wildlife refuge could be established. A technical solution would be to build bigger and better beach stabilization structures. Even though over the long term, these structures may do more harm than good, in the short term they protect the coast. Thus the technological fix is to maintain the beach artificially with a continuous sequence of construction projects.

12. (a) 0 cm; (b) 35 cm; (c) 287 cm; (d) 1432 cm (with all distances measured along the surface of the beach).

Selected Bibliography

Two oceanography texts are:
Paul Pinet: Oceanography, An Introduction to the Planet Oceanus. St.Paul, MN, 1992. 551 pp.

Harold V. Thurman: Intorductory Oceanography. 6th ed. New York, Macmillan Publishing, 1991. 514 pp.

Fundamental ocean processes:
Joan Brown and others: Ocean Chemistry and Deep-sea Sediments. Elmsford, NY, Pergamon, 1989. 134 pp.

Joan Brown and others: Ocean Circulation. Elmsford, NY, Pergamon, 1989. 238 pp.

Joan Brown and others: Waves, Tides, and Shallow-water Processes. Elmsford, NY, Pergamon, 1989. 187 pp.

Robert E. Sheridan and John A. Grow, eds.: The Atlantic Continental Margin. Boulder, CO, Geological Society of America, 1988. 610 pp.

Coastline erosion:
Robert Dolan, Michael Fenster, and Stuart Holme: Erosion of U.S. Coastlines. Geotimes, (23), 1990.

Dennis W. Ducsik: Shoreline for the Public. Cambridge, MA, Massachusetts Institute of Technology Press, 1974. 257 pp.

J. D. Hansom: Coasts. Cambridge, University of Cambridge Press, 1988. 92 pp.

Joseph M. Heikoff: Politics of Shore Erosion: Westhampton Beach. Ann Arbor, MI, Ann Arbor Science Publishers, Inc., 1976. 173 pp.

Thomas C. Jackson and Diana Riesche (eds.): Coast Alert, Scientists Speak Out. San Francisco, The Coast Alliance, 1981. 181 pp.

Jodi L. Jacobson: Swept Away. World Watch, (20), January 1989.

Wallace Kaufman and Orrin H. Pilkey, Jr.: The Beaches Are Moving, The Drowning of America's Shoreline. Durham, N.C., Duke University Press, 1983. 336 pp.

Joseph T. Kelley, Alice R. Kelley, Orrin H. Pilkey, Sr., and Albert A. Clark: Living with the Louisiana Shore. Durham, N.C., Duke University Press, 1984. 164 pp.

Larry R. McCormick, Orrin H. Pilkey, Jr., William J. Neal, and Orrin H. Pilkey, Sr.: Living with Long Island's South Shore. Durham, N.C., Duke University Press, 1984. 157 pp.

Reefs:
Jeremy Stafford-Deitsch: A Safari Through the Coral World. San Francisco, CA, Sierra Club Books, 1991. 200 pp.

Ocean pollution:
Art Davidson: In the Wake of the Exxon Valdez. San Francisco, Sierra Club Books, 1990. 333 pp.

Tom Horton and William M. Eichbaum: Turning the Tide: Saving Chesapeake Bay. Washington, DC, Iskand Press, 1991. 327 pp.

Chapter 13

Seawater and Coastlines

Multiple Choice:

1. Oceans cover about ___ percent of the Earth's surface.
(a) 20; (b) 71; (c) 10; (d) 100; (e) 50

2. The largest and deepest ocean is the
(a) Pacific; (b) Atlantic; (c) Caspian; (d) Arctic; (e) Indian.

3. Seawater contains
(a) trace elements; (b) salts; (c) dissolved oxygen; (d) dissolved carbon dioxide; (e) all of these.

4. The salinity of the oceans has been constant because
(a) salt has been removed from seawater at the same rate at which it has been added; (b) All the Earth's water is the salty; (c) no salt is added; (d) a and b; (e) no salt is removed.

5. Which of these factors influences the size of waves?
(a) fetch; (b) water depth; (c) wind speed; (d) length of time wind has been blowing; (e) all of these.

6. Ocean waves steepen near shore because
(a) they speed up; (b) the wave length increases; (c) friction slows the base; (d) the amplitude decreases;
(e) all of these.

7. Storm winds can push enough water against the shore to raise sea level in a
(a) tsunami; (b) tidal wave; (c) wave crest; (d) fetch; (e) storm surge.

8. Tides are caused by
(a) the Moon; (b) the Sun; (c) ocean currents; (d) the gravitational pulls of the Sun and Moon; (e) none of these.

9. Mid-ocean currents
(a) are caused by prevailing winds; (b) are deflected by the rotation of the Earth; (c) affect climate by moving warm or cold bodies of water; (d) include, for example, the Gulf Stream; (e) all of these.

10. Nutrients are brought from the deep ocean to the surface by
(a) upwellings; (b) tidal waves; (c) storm surges; (d) fetch; (e) wave crests.

11. Deep-sea currents are caused by

(a) prevailing winds; (b) the Earth's rotation; (c) climate; (d) differences in water density; (e) the Coriolis effect.

12. Reefs develop in
(a) shallow, tropical seas; (b) silty places; (c) cold oceans; (d) areas with little sunlight; (e) none of these.

13. Atolls are found in deep water because
(a) corals and reef-building organisms always live in deep water; (b) reef organisms live near the surface and their remains fall to the sea floor when they die; (c) islands cannot sink; (d) reefs grow upward as islands sink; (e) b and d.

14. The foreshore is also called the
(a) high tide mark; (b) beach; (c) intertidal zone; (d) storm surge zone; (e) low tide mark.

15. Most of the sediment found along a coast is carried parallel with the coast by
(a) tsunamis; (b) storm surges; (c) longshore currents; (d) mid ocean currents; (e) undertow.

16. Emergent coastlines
(a) are rich in sediment; (b) are characterized by sea stacks and sea cliffs; (c) form where sea level rises; (d) have deep fjords; (e) all except c.

17. Submergent coastlines
(a) are rich in sediment; (b) form where sea level falls; (c) form where sea level rises or coastal land sinks; (d) form where a coastline has risen; (e) all except c.

18. Cliffs, sea arches, sea stacks, and wave-cut platforms are found on
(a) sediment-rich coastlines; (b) sediment-poor coastlines; (c) submergent coastlines; (d) emergent coastlines; (e) b and c.

19. The two essential ingredients for barrier island formation are
(a) a lack of sediment and waves or currents to transport the sediment away from the coast; (b) a large supply of sediment and waves or currents to transport the sediment along the coast; (c) a submergent coastline and offshore currents; (d) groins and stabilized beaches; (e) none of these.

20. Chesapeake bay is an example of
(a) a tombolo; (b) an estuary; (c) a baymouth bar; (d) fjord; (e) none of these.

21. A/an ______ is built from the shore out into the water to

intercept the steady flow of sand.
(a) groin; (b) barrier island; (c) tombolo; (d) atoll; (e) sea wall.

22. During the past 40,000 years, sea level has fluctuated by (a) 10 meters; (b) 1000 meters; (c) 140 meters; (d) 10 centimeters; (e) only a slight amount.

True or False:

1. As the Earth formed, large amounts of water remaining in the inner Solar System were attracted by the Earth's gravity.

2. The surface of the Arctic Ocean freezes in winter, and parts of it melt for a few months during summer and early fall.

3. The salinity of the oceans is increasing because salt is added continuously by the world's rivers.

4. The movement of water in an ocean wave continues downward below the wave trough in circles of increasing size.

5. The destructive power of a tsunami develops when the wavelength lengthens and the wave height decreases as the wave rolls into shallow water and breaks.

6. Tides occur at the same time each day.

7. An ocean wave moves along the sea surface, but the water in the wave travels in circles, ending up where it started.

8. North-south currents always veer to the right in the Northern Hemisphere because of deflection caused by the Coriolis effect.

9. An individual water molecule that sinks near Greenland may travel for 500 to 2000 years before resurfacing half a world away in the south polar sea.

10. Most sediment found on a coast is produced by erosion at that location.

11. Reef building corals could be adversely affected by a rise in temperature of seawater.

12. Rivers carry sand, silt, and clay to the sea and deposit the sediment on deltas that may cover thousands of square kilometers.

13. Waves inside a bay between two head lands will have higher energy than the waves striking the headland.

14. Continental interiors are generally cooler in summer and warmer in winter than coastal regions.

15. Sea level has risen and fallen repeatedly in the geologic past, and coastlines have emerged and submerged throughout the history of the planet.

16. During the last century, global sea level has not risen.

Completion:

1. The layer below the warm surface layer of the ocean where the temperature drops rapidly with depth is called the ______.

2. The ______ ______ is the vertical distance from the crest to the trough.

3. ______ is the bending of a wave as it strikes shore at an oblique angle.

4. ______ are the vertical displacements of sea level that occur daily.

5. When the both Sun and Moon are directly in line with Earth, their gravitational fields combine, to create a/an ______ ______.

6. Most mid-ocean surface currents move in circular paths called ______.

7. The deflection of currents caused by the Earth's rotation is called the ______ ______.

8. A/an ______ ______ is the flow of ocean water caused by tides.

9. Water flowing back toward the sea after a wave breaks cause a/an ______ current.

10. A/an ______ is a wave-resistant ridge or mound built by corals, algae, and other organisms.

11. A/an ______ is any strip of shoreline washed by waves and tides.

12. If a coastline rises or sea level falls a/an ______ coastline is created.

13. A long ridge of sand or gravel extending out from a beach is called a/an ______.

14. If waves cut a cave into a narrow headland, the cave may eventually erode all the way through the headland, forming a/an ______ ______.

15. A/an ______ is a deep, long, narrow arm of the sea surrounded by high rocky cliffs or mountainous slopes.

16. Both tectonic events and climate change lead to changes in ______ ______.

Answers for Chapter 13

Multiple Choice: 1. b; 2. a; 3. e; 4. b; 5. e; 6. c; 7. e; 8. d; 9. e; 10. a; 11. d; 12. a; 13. d; 14. c; 15. c; 16. a; 17. c; 18. e; 19. b; 20. b; 21. a; 22. c

True or False: 1. T; 2. T; 3. F; 4. F; 5. F; 6. F; 7. T; 8. T; 9. T; 10. F; 11. T; 12. T; 13. F; 14. F; 15. T; 16. F

Completion: 1. thermocline; 2. wave height; 3. refraction; 4. tides; 5. spring tide; 6. gyres; 7. coriolis effect; 8. tidal current; 9. rip; 10. reef; 11. beach; 12. emergent; 13. spit; 14. sea arch; 15. fjord; 16. sea level

CHAPTER 14

The Geology of the Ocean Floor

Discussion

In this chapter, plate tectonics concepts described earlier in Chapter 5 are used to explain the major topographic and bathymetric features of the Earth's ocean basins. In turn, the geology of the sea floor explains and reinforces the plate tectonics model. Figure 14-5, Marie Tharp's sea floor map, is the key figure, showing the bathymetric features of the ocean basins.

The mid-oceanic ridge system is described as a world-encircling spreading center where new lithosphere forms. Trenches are depressions in the sea floor where old lithosphere is consumed back into the mantle at the same rate by subduction. Active and passive continental margins are integrated into the plate tectonics theory. Finally, abyssal plains, submarine canyons, abyssal fans, island arcs, seamounts, and other bathymetric features of the sea floor are described and their respective origins explained.

Chapter 14 contains a Earth Science and the Environment box on Life in the Sea.

Answers to Discussion Questions

1. An oceanic trench forms where old, cold and relatively dense lithosphere sinks beneath warmer, less dense oceanic lithosphere, or beneath continental lithosphere. The subducting plate probably sinks because it has cooled to the point that its density is greater than that of the underlying asthenosphere, and it is therefore out of isostatic equilibrium. A trench forms because the sinking plate frictionally drags the edge of the adjacent plate downward as it sinks. A trench can only form if the rate of sinking exceeds the rate of sediment accumulation in the depression.

2. The east coast of South America is a passive margin. Hence it has accumulated sediment to construct the shelf since the time it rifted from western Africa. In contrast, the western coast of South America is an active margin. Much of the sediment shed into the Pacific coast slides into the trench, and therefore does not accumulate on a stable margin to construct a broad shelf.

3. The junction between continental and oceanic crust is probably characterized by numerous normal faults, or half-grabens, in both the granitic and basaltic sides of the junction. They must have formed during rifting of continental crust, and continued to form as oceanic crust was added to the widening rift. The continental basement rock at the junction may be of any age, but the oceanic crust must have formed at the time rifting was initiated. Arkosic sandstones are probably common near the junction as a result of active normal faulting during rifting. Evaporites are probably also common, from restricted seas associated with a newly opening marine basin with poorly developed access to the open ocean.

The basaltic side of the junction probably is no thicker that normal oceanic crust. The granitic continental crust may thin to only about 10 kilometers as a result of erosion and normal faulting. In places, basalt flows must cover granitic crust as a result of volcanism associated with the beginning of rifting of the continent.

4. Most weathering, erosion, and sediment transport is caused by water. If there were no water, surface features would be altered by wind and mass wasting but rates of weathering and erosion would be much slower than at present. As a result, mountain ranges would be higher and more plateau-like, valleys and canyons less pronounced, and less sediment would accumulate in intracratonic basins and continental margins.

If the entire surface were covered with water, there would still be differentiation into thick granitic and thinner basaltic crust. Surface erosion from streams and glaciers would not occur, so again the shape of mountain ranges and plains would be different.

Selected Bibliography

Several of the references listed for Chapter 5 are also relevant for this chapter.

Enrico Bonatti and Kathleen Crane: Oceanic Fracture Zones. Scientific American, 5: 40-51, 1984.

J. M. Coleman and D. B. Prior: Mass Wasting on Continental Margins. Annual Review of Earth and Planetary Sciences 16 (101), 1988.

Patricia Fryer: Mud Volcanoes of the Marianas. Scientific American, 2: 46-52, 1992.

Chapter 14

Kenneth C. Macdonald and Paul J. Fox: The Mid-Ocean Ridge. Scientific American, 6: 72-79, 1990.

Kenneth C. Macdonald, Daniel Scheirer, and Suzanne M. Carbotte: Mid-Ocean Ridges: Discontinuities, Segments, and Giant Cracks. Science, 253: 986- 994, 1991.

Chapter 14

The Geology of the Ocean Floor

Multiple Choice:

1. The average depth of the sea is
(a) 10 km; (b) 5 km; (c) 350 m; (d) 2 km; (e) 3-4 miles.

2. Samples can be taken from the deep sea floor by
(a) seismic profilers; (b) echo sounders; (c) coring devices; (d) remote sensing; (e) all of these.

3. The mid-oceanic ridge is made of
(a) granite; (b) ocean floor sediments; (c) folded and faulted sedimentary rocks; (d) basalt; (e) none of these.

4. The mid-oceanic ridge rises high above the surrounding sea floor because
(a) new lithosphere forming at the ridge axis is hot; (b) new lithosphere forming at the ridge axis is of relatively low density; (c) new lithosphere forming at the ridge axis of relatively high density (d) a and b; (e) a and c.

5. The highest heat flow on the sea floor is
(a) along the mid-oceanic ridge; (b) on the abyssal plain; (c) on active transforms; (d) on dormant sea mounts; (e) where the lithosphere is thickest.

6. Oceanic crust is relatively young because
(a) it forms continuously at spreading centers and recycles into the mantle at subduction zones; (b) it cannot return to the mantle; (c) it sinks into the mantle at the rift valley; (d) it is made of basalt which is the youngest rock on Earth; (e) none of these.

7. The part of oceanic crust made up of sediment is
(a) Layer 1; (b) Layer 2; (c) Layer 3; (d) under the basalt; (e) the mafic zone.

8. Most of Layer 2 of the oceanic crust is
(a) pelagic sediment; (b) pillow lava; (c) gabbro; (d) terrigenous sediment; (e) basaltic dikes.

9. The gray and red-brown mixture of clay and the remains of tiny plants and animals that live in the surface waters of the oceans is called
(a) zooplankton; (b) terrigenous sediment; (c) pelagic sediment; (d) pillow basalt; (e) gabbro.

10. The flattest surfaces on Earth are
(a) cratons; (b) mid-ocean ridges; (c) abyssal plains; (d) high plains; (e) the continental shelf.

11. The dissolved ions in manganese nodules precipitate by a chemical reaction
(a) between layer 2 and 3; (b) between seawater and terrigenous sediment; (c) between seawater and basalt; (d) between layer 1 and 2; (e) between seawater and pelagic sediment.

12. Seawater heated near _______ and ______ dissolves metals in oceanic crust which then precipitate as huge ore deposits.
(a) the mid-oceanic ridge and submarine volcanoes; (b) atolls and reefs; (c) sea stacks and arches; (d) abyssal plains and a passive continental margin; (e) none of these.

13. A firmly connected boundary between continental and oceanic crust is a
(a) mid-oceanic ridge; (b) passive continental margin; (c) active continental margin; (d) plate boundary; (e) continental shelf.

14. The continental shelf-slope-rise complex on a passive continental margin is
(a) a rugged surface formed by erosion of sediment near the continental margin; (b) a totally flat surface similar to an abyssal plain; (c) a topographic surface formed by accumulation of sediment near the continental margin; (d) a steep surface formed by accumulation of basalt near the continental margin; (e) none of these.

15. Submarine canyons on continental shelves and slopes are cut by
(a) black smokers; (b) longshore currents; (c) deepsea currents; (d) continental rivers and streams; (e) turbidity currents.

16. An active continental margin forms
(a) at mid-oceanic ridges; (b) at continent-continent margins; (c) where an oceanic plate subducts beneath a continental plate; (d) in rift valleys; (e) at black smokers.

17. An active continental margin commonly has a much narrower continental shelf and a considerably steeper continental slope than does a passive margin because
(a) sediment flows into the trench instead of accumulating on the ocean floor; (b) they are younger; (c) they are older; (d) there are no turbidity currents; (e) none of these.

18. When two ocean plates collide one plate
(a) moves away from the other; (b) subducts under the other; (c) dives into the mantle; (d) is partially melted to form magma; (e)

b, c, and d; (f) a, c, and d.

19. Aseismic ridges
(a) are part of the mid-oceanic ridge; (b) form at subduction zones; (c) have no earthquake activity; (d) have frequent earthquake activity; (e) b and c.

20. Seamounts
(a) sink as they become older; (b) sink as they become younger; (c) become less dense as they become older; (d) become hotter as they become older.

True or False:

1. At present, the Atlantic Ocean is shrinking while the Pacific is growing.

2. Remote sensing methods require direct physical contact with the ocean floor.

3. Shallow earthquakes and high heat flow occur at the mid-oceanic ridge.

4. Pillow lavas, basalt dikes, and gabbro sills form only at the mid-oceanic ridge.

5. The extraordinarily level surfaces of the abyssal plains result from accumulation of sediment in the deep ocean.

6. Manganese nodules are found both on the surface of and within ocean floor sediments.

7. Hot seawater picks up metals as it circulates through oceanic crust.

8. A continental shelf on a passive margin is always a small feature.

9. Some of the world's richest offshore petroleum reserves are found on continental shelves.

10. A turbidity current can travel at speeds greater than 100 kilometers per hour and for distances up to 700 kilometers.

11. The deepest place on Earth is in the Mariana trench, north of New Guinea.

12. Subduction at an active continental margin causes no tectonic activity.

13. After a volcanic island or seamount forms, it begins to rise.

14. Seamounts and oceanic islands occur most commonly as isolated peaks on the ocean floor.

15. Earthquakes are common at the mid-oceanic ridge and island arcs.

16. If the Pacific Ocean plate continues to move at its present rate, the island of Hawaii may sink beneath the sea within 10 to 15 million years.

Completion:

1. A/an ______ ______ is an open-mouthed steel net dragged along the sea floor behind a research ship.

2. The ______ ______is the principal tool for mapping sea-floor topography.

3. The ______ ______ is a continuous submarine mountain chain that encircles the globe.

4. Oceanic crust is ______ and ______ at the mid-oceanic ridge and becomes ______ and ______ with increasing distance from the ridge.

5. ______ ______ is composed of sand, silt, and clay eroded from the continents and deposited on the ocean floor near the continents by submarine currents.

6. ______ ______ are jets of black metal-bearing solutions that spout from fractures in the mid-oceanic ridge.

7. Subduction of an oceanic plate beneath a continental plate occurs at a/an______ ______ ______.

8. A/an ______ ______ is a shallow, gently sloping submarine surface on the submerged edge of a continent.

9. A steep region of the sea floor that averages about 50 kilometers wide is the ______ ______.

10. Deep, V-shaped, steep-walled valleys called ______ ______ are eroded into continental shelves and slopes.

11. Turbidity currents slow down when they reach the level deep-sea floor beyond the continental slope where the form ______ ______.

12. A long, narrow, steep-sided depression called a/an ______ forms on the sea floor where the oceanic plate bends downward and begins to subduct.

13. A chain of submarine volcanoes forms at a mid-oceanic trench is a/an ______ ______.

14. A/an ______ is a submarine mountain that rises 1 kilometer or more above the surrounding sea floor.

15. The Hawaiian Island-Emperor Seamount Chain is an example of a/an ______ ______.

Answers for Chapter 14

Multiple Choice: 1. b; 2. c; 3. d; 4. d; 5. a; 6. a; 7. a; 8. b; 9. c; 10. c; 11. e; 12. a; 13. b; 14. c; 15. e; 16. c; 17. a; 18. e; 19. e; 20. a

True or False: 1. F; 2. F; 3. T; 4. T; 5. T; 6. F; 7. T; 8. F; 9. T; 10. T; 11. T; 12. F; 13. F; 14. T; 15. T; 16. T

Completion: 1. rock dredge; 2. echo sounder; 3. mid-oceanic ridge; 4. youngest and of low density; older and denser; 5. terrigenous sediment; 6. black smokers; 7. active continental margin; 8. continental shelf; 9. continental slope; 10. submarine canyons; 11. abyssal fans; 12. trench; 13. island arc; 14. seamount; 15. aseismic ridge

CHAPTER 15

The Earth's Atmosphere

Discussion

The Earth's Atmosphere is the first of four chapters devoted to Atmospheric processes. Chapter 15 discusses the composition and structure of the atmosphere, and the remaining 3 chapters explain the fundamentals of weather, climate, and air pollution. We emphasize climate, climate change, and air pollution becuase these topics play an important role in modern Earth Science.

We open the chapter with a discussion of the evolution of the atmosphere. If students understand that the Earth's atmosphere is different from that of its neighbors, Venus and Mar, and if they study the past interactions between living organisms and atmospheric composition, they will better appreciate present and future concerns such as the depletion of stratospheric ozone and the greenhouse effect.

We discuss radiation and energy transfer in the atmosphere as a prelude to the greenhouse effect introduced here and in Chapter 17. These sections are followed with discussions of temperature changes with elevation and latitude and with an introduction to heat transport and storage.

Depletion of the stratospheric ozone layer is discussed in an Earth Science and the Environment box. Chapter 15 also contains an Earth Science and the Environment box on solar cells. In addition two Focus On boxes: The Upper Fringe of the Atmosphere and Latitude and Longitude are included.

Answers to Discussion Questions

1. The Earth's primitive atmosphere was devoid of oxygen. Not only do we need oxygen to breathe, but oxygen converts to ozone in the stratosphere, thus filtering out harmful ultraviolet rays. According to one theory, the ultraviolet rays that would harm life today, and that are filtered out by the present atmosphere, were responsible for initiating the reactions that eventually led to the formation of life.

2. Pressure would decrease and initially the atmosphere would become thicker, ie its volume would increase to accommodate the decrease in pressure. However, with reduced gravity, many of the gases would eventually escape into space, leading to a permanent loss of much of the atmosphere.

3. The Sun emits a wide range of electromagnetic frequencies, including high energy ultraviolet, X-rays, and gamma radiation that are harmful to human beings. Therefore, astronauts must wear protective clothing in outer space. Most of the high energy rays (and many low ones) are absorbed in the upper atmosphere and do not reach the surface of the Earth. Therefore, people on the Earth do not need to wear such protective clothing.

4. There is more high frequency, high energy light on a mountain top than at sea level because many of these rays have not yet been absorbed in their passage through the atmosphere to the surface of the Earth. Sunburn is caused only by more energetic photons, you can sit all day in a warm room in front of a hot stove and not get a trace of a burn.

5. Snow has a high albedo and reflects most of the heat back into the atmosphere. Trees, twigs, and rocks absorb sunlight and become warm, thereby melting adjacent snow.

6. Alaska is so far north that in winter the sunlight strikes the Earth at a low angle even at noon, analogous to sunrise and sunset at lower latitudes.

7. (a) 47 % (b) Although climate changes with time, these changes occur slowly and usually represent only a few degrees variation. Since large amounts of solar radiation are absorbed, and temperature remains relatively constant, heat must be re-radiated from the ground. The Earth absorbs short wavelength visible and UV radiation and re-radiates the same amount of energy at longer wavelengths.

8. Because the hypothetical Earth described in this problem is tilted, different regions would experience different climates. But the differences would not be as great as they are now because the curvature of a spherical Earth tilts the polar regions farther away from a direct line with the Sun's rays.

9. The sunlight at the poles arrives at a low angle and is therefore less concentrated than the sunlight at the Equator.

10. If a lake froze in winter, its albedo and heat storage capacities would become very much like snowy land surfaces, and the lake would have little effect on local climate. If, however, the lake remained unfrozen, its heat transport and storage would be similar to that of the ocean, as explained in the text.

11. If oxygen and nitrogen absorbed visible radiation, the atmosphere would be opaque to sunlight and the upper atmosphere would be extremely hot. Life could not exist on the surface of the Earth. If oxygen and nitrogen absorbed infrared radiation,

it would absorb radiation emitted from the ground and the Earth would become unbearably hot. However, the present atmospheric composition would never have evolved, because living organisms would not have evolved, or if evolution had started, life would have died out as the Earth heated.

Selected Bibliography

In recent years there has been significant debate surrounding the questions: "What was the composition of the Earth's primitive atmosphere and how did life form?" Four references for further study are:

S.K. Atreya, J.B. Pollack, and M.S. Matthews, eds: Origin and Evolution of Planetary and Satellite Atmospheres. Tuscon, University of Arizona Press, 1989. 881 pp.

Stephen L. Gillett: The Rise and Fall of the Early Reducing Atmosphere. Astronomy, 7: 66 ff, 1985.

J. E. Lovelock: Gaia. Oxford, Oxford University Press, 1979. 157 pp.

Stephen Schneider and Randi Londer: The Coevolution of Climate and Life. San Francisco, Sierra Club Books, 1985. 576 pp.

Discussion of the ozone crisis are:

Jack Fishman and Robert Kalish: Global Alert. New York, Plenum Press, 1990. 311 pp.

Sharon A. Roan: Ozone Crisis. New York, Wiley and Sons, 1990. 270 pp.

Chapter 15

The Earth's Atmosphere

Multiple Choice:

1. The Earth's original atmosphere
(a) had oxygen as a major constituent; (b) was an effective filter that absorbed high energy ultraviolet rays; (c) would be poisonous to most forms of life today; (d) was formed by biological processes.

2. The Earth's modern atmosphere
(a) is composed of 80 % oxygen; (b) does not filter high energy ultraviolet rays; (c) is similar to that found on Venus; (d) is composed of 78 % nitrogen.

3. As early organisms evolved and multiplied, they released more ______ into the atmosphere.
(a) oxygen; (b) hydrogen; (c) nitrogen; (d) argon; (e) glucose

4. The presence of appreciable amounts of oxygen in our atmosphere can be explained by
(a) meteorite impacts; (b) biological activity; (c) tectonic activity; (d) the big bang theory.

5. The gaseous composition of dry air is
(a) mostly nitrogen; (b) mostly oxygen; (c) mostly helium; (d) low in nitrogen.

6. The atmosphere
(a) becomes more dense with increasing altitude; (b) becomes less dense with increasing altitude; (c) remains at constant density with increasing altitude; (d) becomes less dense throughout the troposphere and then more dense in the stratosphere.

7. In a barometer the weight of the atmosphere, is compared with the weight of
(a) an equal column of nitrogen; (b) an equal column of oxygen; (c) a densely packed cylinder; (d) a vacuum.

8. Which is not an electromagnetic wave?
(a) light; (b) sound; (c) radio waves; (d) X-rays; (e) infrared rays

9. Radiation with long wavelength and low frequency
(a) is high energy radiation; (b) is low energy radiation; (c) has the same energy as all other forms of radiation; (d) is called X-rays.

10. When an object absorbs radiation,
(a) the photons disappear and convert to another form of energy; (b) the energy is lost; (c) the photons reflect out to space; (d) the photons slow down; (e) the photons speed up.

11. The sky appears blue because
(a) Earth's atmosphere scatters blue and violet wavelengths through the sky; (b) light is reflected by the Earth's oceans; (c) there is no atmosphere to scatter the light in the uppermost atmosphere; (d) only blue frequencies are transmitted by the Sun; (e) none of these.

12. If scientists were to spread a thin layer of ashes on the Polar Ice Cap, we would expect that the albedo of the region would
(a) increase leading to a warming of the region; (b) increase leading to a cooling; (c) decrease leading to a warming; (d) decrease leading to a cooling.

13. ______ percent of the incoming solar radiation reaches Earth's surface.
(a) 1 % (b) 99 %; (c) 8 %; (d) 47%; (e) 80%

14. Why doesn't Earth get hot enough to boil oceans and melt rocks from the radiant energy it absorbs from the Sun?
(a) The energy disappears. (b) The atmosphere cools the Earth. (c) The sun sets at night. (d) The Earth re-emits all the energy it absorbs. (e) The oceans keep the Earth cool.

15. The Earth remains warm at night because
(a) the albedo absorbs and retains much of the radiation emitted by the ground; (b) the atmosphere absorbs and retains much of the radiation emitted by the ground; (c) the atmosphere absorbs and retains much of the radiation emitted by the Moon; (d) dark sky retains more heat than blue sky; (e) none of these.

16. Which of the following orders of the various layers of the atmosphere is correct? (Start with the lowest layer)
(a) thermosphere < stratosphere < troposphere < mesosphere; (b) troposphere < mesosphere < stratosphere < thermosphere; (c) stratosphere < troposphere < mesosphere < thermosphere; (d) troposphere < stratosphere < mesosphere < thermosphere.

17. The temperature of the stratosphere increases with elevation because the stratosphere
(a) is heated primarily by the Earth; (b) is heated primarily by solar radiation; (c) is located closer to the equator; (d) is closer to the Earth's surface; (e) none of these.

18. The region of the globe directly beneath the Sun is warmer because
(a) it receives the most concentrated radiation; (b) it receives the least concentrated radiation; (c) it receives light of a different frequency; (d) it receives light of a different wavelength.

19. Convection readily occurs
(a) only in liquids; (b) only in solids; (c) only in gases; (d) in liquids and solids; (e) in liquids and gases; (f) in solids and gases.

20. ______ plays an important role in heat transfer and storage because it commonly occurs naturally on Earth in all three states, solid, liquid, and gas.
(a) hydrogen; (b) air; (c) water; (d) rock; (e) soil

21. If you place a pan of water and a rock outside on a hot summer day,
(a) the rock becomes hotter than the water; (b) the water becomes hotter than the rock; (c) the rock and the water heat to equal temperatures; (d) the rock receives a larger quantity of solar radiation; (e) none of these

True or False:

1. If you landed on the Moon without a space suit, you would not live long.

2. As early organisms evolved and multiplied, they released more oxygen into the atmosphere.

3. If all life on Earth were to cease, the atmosphere would revert to its primitive, oxygen-poor composition and become poisonous to modern plants and animals.

4. The amount of radiation that the Earth receives from the Sun is so small that it hardly warms the Earth's surface.

5. Photons appear when they are emitted and disappear when they are absorbed.

6. Visible light is a large portion of the electromagnetic spectrum.

7. Radiation can travel through space for trillions upon trillions of kilometers without any change in wavelength or loss of energy.

8. The Earth emits higher energy radiation than the Sun does.

9. If Earth had no atmosphere, sunlight would travel directly to the surface and the Sun would look white in a black sky.

10. If the air is filled with dust or water droplets, all the wavelengths scatter and the sky becomes red.

11. If the Earth's albedo were to rise by growth of glaciers or increase in cloud cover, the surface of our planet would warm.

12. If Earth had no atmosphere, radiant heat loss would be so rapid that the Earth's surface would cool drastically at night.

13. The ozone in the upper atmosphere protects life on Earth by absorbing much of the high-energy radiation before it reaches Earth's surface.

14. Even though the temperature in the thermosphere is no colder than that on a winter day in Chicago, if you ascended into the thermosphere you would rapidly freeze to death.

15. All areas of the globe receive the same total number of hours of sunlight every year.

16. Air is a good conductor of heat.

17. Continental interiors are commonly cooler in summer and warmer in winter than coastal areas.

Completion:

1. ______ is the state of the atmosphere at a given place and time.

2. ______ is a composite of weather patterns from season to season, averaged over many years.

3. Fires burn rapidly if ______ is abundant, so if its concentration were to increase even by a few percent, fires would burn uncontrollably across the planet.

4. The pressure exerted by the air is the ______ ______.

5. Particles of light are called ______.

6. The ______ ______ is the continuum of radiation of different wavelengths.

7. The Earth re-emits ______ radiation, which has relatively long wavelength and low frequency.

8. Atmospheric gases, water droplets, and dust particles ______ sunlight in all directions.

9. ______ is the reflectivity of a surface.

10. The ______ traps heat radiating from Earth and acts as an insulating blanket.

11. The layer of air closest to the Earth, the layer we live in, is the ______.

12. Little radiation is absorbed in the ______, and the thin air is extremely cold.

13. On the ______ every portion of the globe receives 12 hours of sunlight and 12 hours of darkness.

14. ______ is the transport of heat by collisions between atoms and molecules.

15. ______ is the transport of heat by currents in liquids or gases.

16. ______ ______ is the energy released or absorbed when a substance changes from one state to another.

17. ______ ______ is the amount of energy needed to raise the temperature of 1 gram of material by 1° C.

Answers for Chapter 15

Multiple Choice: 1. c; 2. d; 3. a; 4. b; 5. a; 6. b; 7. d; 8. b; 9. b; 10. a; 11. a; 12. c; 13. d; 14. d; 15. b; 16. d; 17. b; 18. a; 19. e; 20. c; 21. a

True of False: 1. T; 2. T; 3. T; 4. F; 5. T; 6. F; 7. T; 8. F; 9. T; 10. F; 11. F; 12. T; 13. T; 14. T; 15. T; 16. F; 17. F

Completion: 1. weather; 2. climate; 3. oxygen; 4. barometric pressure; 5. photons; 6. electromagnetic spectrum; 7. infrared; 8. scatter; 9. albedo; 10. atmosphere; 11. troposphere; 12. mesosphere; 13. equinox; 14. conduction; 15. convection; 16. latent heat; 17. specific heat

CHAPTER 16

Weather

Discussion

Most students watch weather reports on TV and have learned generalizations, such as low pressure systems bring rain and high pressure brings sunshine. They have heard terms such as cold fronts, warm fronts, and jet stream and have watched the sky as stratus clouds bring steady rain or cumulus clouds build on a summer afternoon to towering thunderheads. In this chapter we can provide scientific background to explain these relationships.

The weather is defined primarily by moisture and wind. Therefore we start the chapter with a discussion of humidity, clouds, and precipitation, and follow with pressure and wind. With this background, the student can understand the winds and rainfall that occur when air masses collide, creating frontal weather systems. The chapter ends with a discussion of violent storms: thunderstorms, tornadoes, and tropical cyclones.

Chapter 16 contains an Earth Science and the Environment box on cloud seeding and a Memory Device box on names of clouds.

Answers to Discussion Questions

1.

Temperature	Amount of water when air is saturated (g/m^3)	Amount of water at 50 % relative humidity (g/m^3)
0° C	4.8	2.4
10° C	9.4	4.7
20° C	17.3	8.6
40° C	51.2	25.6

2. Frost forms when vapor-laden air inside the refrigerator comes in contact with the cooling coils. More frost would form in summer in a humid region where the outside air is warm and moist.

3. The most condensation occurs when the maximum amount of moisture evaporates into the air during the day and the temperature difference between night and day is greatest.

Condition (a) would not lead to condensation, (b) would produce dew, (c) would produce frost.

4. (a) Clouds form at 1900 m. (b) The air mass is the same temperature as ambient air so it never rises and no clouds form. (c) Clouds form at 1900 m [see (a), above]. (d) Air mass stops rising before condensation occurs.

6. Wind is powered by the Sun's energy. Thus wind is a heat engine and obeys the same thermodynamic laws as any heat engine such as an automobile engine or a steam fired turbine.

7. A sunny day because the Sun warms the Earth creating temperature differences between ground and sea.

8. As explained in the text, such a collision produces low pressure and rain. The characteristics of the storm depend on the relative velocities, temperature, and moisture content of the two air masses.

Selected Bibliography

Four books on weather systems and a timely article on weather are:

W.J. Burroughs: Watching the World's Weather. Cambridge, Cambridge University Press, 1991. 196 pp.

David M. Ludlum: The Audubon Society Field Guide to North American Weather. New York, Knopf, 1991. 656 pp.

J.F.R. McIlveen: Fundamentals of Weather and Climate. London, Cahpman and Hall, 1991. 544 pp.

Leslie Musk: Weather Systems. Cambridge, Cambridge University Press, 1988. 158 pp.

Earle R. Williams: The Electrification of Thunderstorms. Scientific American, 11: 88-99, 1988.

Chapter 16

Weather

Multiple Choice:

1. Which of the following statements about humidity is correct? (a) Cold air can hold more moisture than warm air. (b) Rain or fog is likely to occur when cool moist air is heated. (c) Dew is forms when warm moist air comes in contact with cool surfaces. (d) Fog occurs when warm moist air from the ocean blows over warmer land surfaces.

2. ______ ______ is the amount of water vapor in air relative to the maximum it can hold at a given temperature.
(a) Relative humidity; (b) absolute humidity; (c) dew point; (d) adiabatic lapse rate; (e) partial humidity

3. Moisture condenses from air when the air
(a) loses heat by radiation; (b) contacts cool surfaces; (c) rises; (d) a, b, and c; (e) a and b; (f) a and c.

4. Which of the following processes would lower the relative humidity of a parcel of air?
(a) The air moves over a lake where water is evaporating. (b) The air is heated and no water enters or leaves the system. (c) The air is cooled and no water enters or leaves. (d) Dust blows into the air but no other changes occur.

5. Which of the following conditions is likely to lead to rain?
(a) warm, moist air rises; (b) cool, moist air falls; (c) warm air makes contact with cool surfaces; (d) moist air sinks as it passes over the lee side of a mountain

6. Supercooled droplets do not freeze because
(a) there are too many particles to condense onto; (b) they are not cold enough; (c) there are no solid particles to condense onto; (d) they have a natural antifreeze in them; (e) none of these.

7. During an adiabatic temperature change
(a) air is heated by the sun; (b) air is cooled by the polar ice caps; (c) a and b; (d) air cools by evaporation; (e) the temperature rises or falls with gain or loss of heat.

8. Once clouds start to form, rising air
(a) cools more quickly than it did lower in the atmosphere; (b) heats up; (c) no longer cools as rapidly as it did lower in the atmosphere; (d) turns into rain; (e) none of these.

9. If air stops rising before it cools to its dew point,
(a) clouds form; (b) no clouds form; (c) it rains; (d) it snows; (e) none of these.

10. When condensation occurs at the same elevation at which air stops rising,
(a) cumulus clouds form; (b) cirrus clouds form; (c) stratus clouds form; (d) no clouds form; (e) all of these.

11. Rain doesn't fall from all clouds because
(a) some contain water droplets that evaporate before they reach Earth; (b) some contain no water droplets; (c) some contain water droplets that are too large to fall; (d) none of these.

12. If one portion of the atmosphere becomes warmer than the surrounding air,
(a) the warm air expands and rises; (b) a high pressure forms; (c) clouds disappear; (d) the warm air contracts and falls; (e) none of these.

13. Sinking air that exerts a downward force forms a
(a) low-pressure region; (b) high-pressure region; (c) rainstorm; (d) hurricane; (e) orographic region.

14. Since warm air can hold more moisture than cold air, clouds generally do not form over
(a) low-pressure regions; (b) high-pressure regions; (c) mountains; (d) the seashore; (e) none of these.

15. If a large pressure difference occurs over a short distance
(a) there is little wind; (b) it rains; (c) it snows; (d) wind blows rapidly; (e) wind blows slowly.

16. A high altitude wind system traveling south from the Equator toward the South Pole
(a) would not be affected by the Earth's spin, because only surface winds are deflected; (b) would be deflected to the east by the Earth's spin; (c) would be deflected to the west by the Earth's spin; (d) would be accelerated southward by the Earth's spin.

17. Orographic lifting creates abundant precipitation on the
(a) leeward side of a mountain; (b) windward side of a mountain; (c) on mountain crests; (d) b and c; (e) in the rain shadow.

18. When a warm air mass overtakes a cool one
(a) the cool air rises, leading to precipitation; (b) the cool air rises, leading to fair weather; (c) the warm air rises, leading to precipitation; (d) the warm air rises, leading to fair weather.

19. Monsoons
(a) blow from land to sea in early summer bringing drought; (b) blow from sea to land when the Earth cools, bringing rain; (c) blow from sea to land in summer bringing rain; (d) blow from land to sea in summer bringing rain.

20. Tropical cyclones
(a) form only over warm oceans; (b) form over any ocean; (c) require warm, moist air; (d) can be detected by satellites; (e) a, c, and d

True or False:

1. Warm air can hold less water vapor than cold air.

2. Rain is almost always caused by cooling that occurs when air rises.

3. If dry air were to rise from sea level to 9000 meters (about the height of Mount Everest), it would cool by 90° C (162° F).

4. Rising air warms adiabatically.

5. Evaporation fogs are common in late fall and early winter, when the air has become cool but the water is still warm.

6. About one million cloud droplets must combine to form an average-size raindrop.

7. Hail always falls from a stratus cloud.

8. When a mass of cool air comes in contact with warm air the cool air rises over the warm air.

9. Wind always flows away from a region of high pressure and toward a low-pressure region.

10. Pressure-gradients change slowly, once or twice a year.

11. The jet stream caused U.S. bombers to miss their targets on the first mass bombing of Tokyo during World War II.

12. Windward valleys receive much less moisture than leeward valleys.

13. When a warm front overtakes a cold front, precipitation is light but widespread.

14. Sea and land breezes are caused by uneven heating and cooling of land and water.

15. A single bolt of lightning produces as much power as a nuclear power plant for a few seconds.

Completion:

1. ______ is the amount of water vapor in air.

2. When relative humidity reaches 100 percent, the air is ______.

3. Air becomes ______ when it has cooled below its dew point but water remains as vapor.

4. A/an ______ is a visible concentration of water droplets or ice crystals in air.

5. ______ are broad sheet-like clouds.

6. ______ clouds are column-like clouds with flat bottoms and a billowy top.

7. ______ ______ occurs when warm moist air from the sea blows onto cooler land.

8. When air blows against a mountainside, it is forced to rise, by a mechanism called ______ ______.

9. ______ is the horizontal movement of air in response to differences in air pressure.

10. A low-pressure region with its accompanying surface wind is called a/an ______.

11. A high-pressure region with outward spiralling surface wind is called a/an ______.

12. A/an ______ ______ forms when a warm air mass is trapped between two colder air masses.

13. When air currents rise and fall simultaneously within the same cloud, ______ ______ is created.

14. A/an ______ is a small, short-lived, funnel-shaped storm that protrudes from the base of a cumulonimbus cloud.

Answers for Chapter 16

Multiple Choice: 1. c; 2. a; 3. d; 4. b; 5. a; 6. c; 7. e; 8. c; 9. b; 10. c; 11. a; 12. a; 13. b; 14. b; 15. d; 16. b; 17. d; 18. c; 19. c; 20. e

True or False: 1. F; 2. T; 3. T; 4. F; 5. T; 6. T; 7. F; 8. F; 9. T; 10. F; 11. T; 12. F; 13. T; 14. T; 15. T

Completion: 1. humidity; 2. saturated; 3. supersaturated; 4. cloud; 5. stratus; 6. cumulus; 7. advection fog; 8. orographic lifting; 9. wind; 10. cyclone; 11. anticyclone; 12. occluded front; 13. wind shear; 14. tornado

CHAPTER 17

Climate

Discussion

The first two sections of this chapter outline the mechanisms that produce global winds and precipitation zones. Major climate zones of the Earth are then described according to the Koeppen climate classification system. Finally, we discuss climate change and human impact on climate such as the greenhouse effect.

Scientists continue to debate whether long term global warming is occurring and if it is, whether human activity is responsible. In this discussion, therefore, we have been careful evaluate the reliability of the data, and to distinguish fact from inference. We hope to encourage classroom discussion of this important topic.

After this manuscript was submitted, an article appeared in Science that has become part of the global warming debate[1]. R. J. Charlson et al argue that air pollutant aerosols cool the earth's surface and that this global cooling is approximately equal to greenhouse warming. We suggest a classroom debate on the question "if these two effects cancel each other, might it be better to allow both emission of greenhouse gases and pollution rather than to bear the expense of reducing both?"

Chapter 17 contains three Earth Science and the Environment boxes: Tropical Rainforests and Climate, Expansion and Contraction, and Threshold Effects.

Answers to Discussion Questions

1. See text.

2. The center of the doldrum area lies in the hottest part of the Earth. Since the Sun shines directly over the equator only during the spring and autumn equinoxes, the doldrums drift north of the Equator in June and July, and south in December and January. Since seasons repeat themselves on a yearly basis, there is less year-to-year variation than month to month variation.

---

[1] R. J. Charlson, S. E. Schwartz, J. M. Hales, R. D. Cess, J. A. Coakley, Jr., J. E. Hansen, and D. J. Hofmann, "Climate Forcing by Anthropogenic Aerosols", Science, 255 (1992): 423.

3. Airline pilots would expect high altitude winds moving from the southwest.

4. The warm air cools as it ascends and eventually stops rising when it hits the tropopause and the warmer air in the stratosphere.

5. (a) Central Oregon which is temperate rainshadow desert or semi-arid praire, (b) Portland Oregon which has a marine west coast climate, (c) Amazon basin rainforest which has a humid tropical climate, (d) Sahara desert, a desert formed by subtropical highs, (e) New Orleans which has humid subtropical climate

7. The urban heat island affect has no appreciable impact on global agriculture. Warmer winters are more comfortable and save heating fuel, but hotter summers are less comfortable and require more energy for air conditioning. Increased fogs and rain mean fewer days with cheery Sun.

8. The major greenhouse gases are: carbon dioxide, methane, chlorofluorocarbons, and oxides of nitrogen. Carbon dioxide is released whenever fuels are burned. Methane is released by bacteria that live in bogs, landfills, rice paddies, and the intestines of grazing animals and termites. It is also released during coal extraction and drilling for petroleum and natural gas. Chlorofluorocarbons are used as spray can propellants, cleaning solvents, refrigerants, and as ingredients in industrial foams. Nitrogen oxides are air pollutants generated chiefly in automobile exhaust.

9. A change in global temperature can have profound impact on agriculture. It may also shift precipitation patterns that affect agriculture. Finally, global warming could affect sea level.

11. Continental interiors are hotter in summer and colder in winter than coastal areas. They are also usually drier. A supercontinent has more interior regions compared to several separated continents. Additionally if continental suturing created Himalayan-like mountain chains, elevated plateaus and rain shadow deserts might be common. Global temperatures would be affected by the latitude of the hypothetical super continent.

Selected Bibliography

A general textbook on climate is:
Michael R. Rampino, John E. Sanders, Walter S. Newman, and L.K. Konigsson, eds: Climate. New York, Van Nostrand Reinhold, 1978. 608 pp.

Six books on global warming are:
Dean Edwin Abrahamson: The Challange of Global Warming. New York, Island Press, 1989. 350 pp.

Thomas Levenson: Ice Time. New York, Harper & Row, 1990. 242 pp.

Francesca Lyman, Irving Mintzer, Kathleen Courrier, and James MacKenzie: The Greenhouse Trap: What We are Doing to the Atmosphere and How We Can Slow Global Warming. Boston, Beacon Press, 1990. 190 pp.

S. Fred Singer: Global Climate Change. New York, Paragon House, 1989. 424 pp.

Steven H. Schneider: Global Warming: Are We Enetering the Greenhouse Century? San Francisco, Sierra Club Books, 1989. 317 pp.

United Nations Environment Program: The Greenhouse Gases. Nairobi, Kenya, UNEP, 1987. 40 pp.

A classic paper on advances and retreats of the Pleistocene glaciers, and a modern sequel are:
J. D. Hays, John Imbrie, and N. J. Shackleton: Variations in the Earth's Orbit: Pacemaker of the Ice Ages. Science, 194:(1121), 1976.

Wallace S. Broecker and George H. Denton: What Drives Glacial Cycles? Scientific American, 1: 49-56, 1990.

A recent, provocative article on human impact on climate is:
R. J. Charlson, S. E. Schwartz, J. M. Hales, R. D. Cess, J. A. Coakley, Jr., J. E. Hansen, and D. J. Hofmann, Climate Forcing by Anthropogenic Aerosols. Science, 225:423, 1992.

Chapter 17

Climate

Multiple Choice:

1. The three-cell model of global wind circulation shows (a) three cells of global air flow bordered by three high-pressure regions; (b) three cells of global air flow bordered by three low-pressure regions; (c) three cells of global air flow bordered by alternating bands of high and low pressure; (d) that global wind systems are generated solely by temperature differences between the equator and the poles; (e) none of these.

2. The horse latitudes (30° north and south latitudes) and the doldrums (approximately 0° latitude) are both noted for regular calms, but they differ in that (a) air in the doldrums is moving vertically downward while air in the horse latitudes is moving vertically upward; (b) rain is frequent in the doldrums while it is much less frequent in the horse latitudes; (c) rain is infrequent in the doldrums while it is more frequent in the horse latitudes; (d) the doldrums is a region of high pressure and the horse latitudes have low pressure.

3. The polar easterlies and prevailing westerlies converge at approximately (a) 60° latitude; (b) 60° longitude; (c) 30° latitude; (d) 0° latitude; (e) the poles.

4. In the three-cell model, (a) air rises in both the horse latitudes and the doldrums; (b) air falls in both the horse latitudes and the doldrums; (c) air rises in the horse latitudes and falls in the doldrums; (d) air rises in the doldrums and falls in the horse latitudes.

5. In the three-cell model, global winds are generated by (a) adiabatic expansion; (b) the Coriolis effect; (c) the jet stream; (d) heat-driven convection currents that are directed by the Earth's rotation; (e) none of these.

6. Which of the following statements about climate is correct? (a) Temperature decreases steadily with increasing longitude. (b) Temperature increases steadily with increasing altitude. (c) Modern climate is affected by frequent meteorite impacts. (d) Seasonal temperature changes are more pronounced at high latitudes than at lower latitudes.

7. The Earth's major climate zones are classified primarily by (a) temperature and precipitation; (b) latitude and altitude; (c)

pressure and wind belts; (d) geographic position; (e) oceans and mountains.

8. The climate zone (or biome) associated with rainfall less than 25 cm/yr is
(a) tropical rainforest; (b) tropical savanna; (c) desert; (d) Mediterranean; (e) marine west coast.

9. The climate zone (or biome) associated with dry summers, rainy winters, and moderate temperature year-round is
(a) tropical rainforest; (b) tropical savanna; (c) desert; (d) Mediterranean; (e) marine west coast.

10. The climate zone (or biome) associated with abundant rainfall year round and greater temperature difference between night and day than from June to December is
(a) tropical rainforest; (b) tropical savanna; (c) desert; (d) Mediterranean; (e) marine west coast.

11. The southeastern United States is an example of a
(a) savannah; (b) tropical grassland; (c) humid subtropical region; (d) polar region; (e) arid subtropical region.

12. Areas that are influenced by warm ocean currents with cool summers, warm winters and abundant precipitation year-round are in the ______ climate zone.
(a) Mediterranean; (b) marine west coast; (c) humid tropical; (d) humid mid-latitude; (e) none of these.

13. As warm air rises over a city, a local low-pressure zone develops, and rainfall is
(a) generally greater over the city than in the surrounding areas; (b) generally less over the city than in the surrounding areas; (c) generally the same over the city as in the surrounding areas; (d) sparse.

14. Tectonic changes
(a) cannot cause climate change; (b) cause climate change only when the continents are near the Poles; (c) may cause climate change over periods of tens of millions of years; (d) may have caused the advances and retreats of the Pleistocene glaciers; (e) cause changes in the Earth's orbit around the Sun.

15. Changes in the Earth's orbit around the Sun
(a) cannot cause climate change; (b) cause climate change only when the continents are near the Poles; (c) may cause climate change over periods of tens of millions of years; (d) may cause climate change over periods of hundreds of millions to a few billion years; (e) may have caused the advances and retreats of the Pleistocene glaciers.

16. A giant meteorite impact would cause climate change
(a) over days to weeks; (b) over thousands of years; (c) over tens of thousands of years; (d) over tens of millions of years; (e) over a few billion years.

17. During photosynthesis
(a) carbon dioxide, which is a greenhouse gas, is released; (b) oxygen, which is a greenhouse gas, is released; (c) carbon dioxide, which is not a greenhouse gas, is released; (d) oxygen, which is not a greenhouse gas, is released.

18. Which of the following changes in energy policy would reduce emission of greenhouse gases?
(a) Switch from high sulfur coal to low sulfur coal. (b) Switch from gasoline to coal. (c) Switch from coal to gasoline. (d) Switch from coal to natural gas. (e) Switch from coal to nuclear fuels.

19. Which of the following is not true for carbon dioxide? It (a) dissolves in seawater; (b) absorbs infrared radiation; (c) reacts with other chemicals to form solid rocks; (d) is destroyed when organic matter is burned.

20. Global atmospheric carbon dioxide has ______ during the last 120 years.
(a) remained the same; (b) increased; (c) decreased; (d) become mostly tied up in limestone; (e) none of these.

True or False:

1. Global wind systems are generated only by the temperature difference between the equator and the poles.

2. High-altitude winds do not continue to flow due north and south because they are deflected by the Coriolis effect.

3. Many of the world's great deserts are located in the horse latitudes.

4. The great wheat belts of the United States, Canada, and Russia all lie between 60° and 90° north latitude.

5. In regions where warm oceans lie adjacent to cooler land, humid maritime air blows inland, leading to abundant precipitation.

6. Visual climate classification is impossible.

7. A tropical monsoon climate has greater total precipitation and greater monthly variation than tropical savanna.

8. In a Mediterranean climate more than 75 percent of the annual rainfall occurs in summer.

9. Temperate rainforests are common along the northwest coast of North America, from Oregon to Alaska.

10. The climate of cities is measurably different from that of the surrounding rural regions.

11. A storm front may remain over a city longer than it would the surrounding country side.

12. No obvious relationship exists between short-term fluctuations in solar output and climate.

13. Milankovitch's calculations showed that the three regular orbital variations should combine periodically to generate alternating cool and warm climates in the higher latitudes.

14. A meteorite impact could produce a cloud that blocks the Sun, and alter climate.

15. It is possible that human actıvities are altering climate, and these alterations may, in turn, affect the human condition and the survival of other species.

Completion:

1. Surface winds moving toward the equator are deflected by the ______ ______, so they blow from the northeast in the Northern Hemisphere and from the southeast in the Southern Hemisphere.

2. The predominant winds in the mid-latitudes are called the ______ ______.

3. Air is forced upward at the convergence of the prevailing westerlies and the polar easterlies, forming a low-pressure boundary zone called the ______ ______.

4. The ______ ______ marks the boundary between cold polar air and the warm, moist westerly flow that originates in the subtropics.

5. The ______ ______ ______ is used by climatologists throughout the world to define principle climate groups.

6. A/an ______ is a community of plants living in a large geographic area characterized by a particular climate.

7. A/an ______ ______ is a grassland with scattered small trees and shrubs.

8. The ______ ______ is characterized by dry summers, rainy winters, and moderate temperature.

9. ______ ______ grow in temperate regions where rainfall is greater that 100 centimeters per year (40 in/yr) and is constant throughout the year.

10. The polar biome where trees cannot survive, and low-lying plants such as mosses, grasses, flowers, and a few small bushes cover the land is called ______.

11. The fact that the center of Washington DC is more than 3° warmer than outlying areas demonstrates the ______ ______ ______ effect.

12. Tectonic events are only responsible for ______ climate change.

13. ______ describes the way the shape of the Earth's orbit around the Sun changes on about a 100,000 year cycle.

14. Seasonal changes in sunlight reaching the Earth can cause an onset of ______ by affecting summer temperature.

15. The warming of the atmosphere caused by an increase in the concentration of infrared absorbing gases is called the ______ ______.

Answers for Chapter 17

Multiple Choice: 1. c; 2. b; 3. a; 4. d; 5. d; 6. d; 7. d; 8. c; 9. d; 10. a; 11. c; 12. b; 13. a; 14. c; 15. e; 16. a; 17. d; 18. e; 19. d; 20. b

True or False: 1. F; 2. T; 3. T; 4. F; 5. T; 6. F; 7. T; 8. F; 9. T; 10. T; 11. T; 12. T; 13. T; 14. T; 15. T

Completion: 1. Coriolis effect; 2. prevailing westerlies; 3. polar front; 4. jet stream; 5. Koeppen climate classification; 6. biome; 7. tropical savanna; 8. Mediterranean climate; 9. temperate rainforests; 10. tundra; 11. urban heat island; 12. long-term; 13. eccentricity; 14. glaciation; 15. greenhouse effect

CHAPTER 18

Air Pollution

Discussion

Destruction of the ozone layer was discussed in Chapter 15 during the explanation of energy transfers in the stratosphere. The greenhouse effect was discussed in Chapter 17 under the heading, climate change. This chapter introduces the major sources of toxic air pollutants and their effects. Two atmospheric mechanisms exacerbate air pollution: secondary chemical reactions that produce smog and acid precipitation, and meteorological conditions that concentrate pollutants in inversion layers. These discussions are followed by a brief review of the effects of air pollution on vegetation, materials, and human health.

In recent years, both scientists and non-scientists have become increasingly aware of and involved in public policy decisions regarding pollution control. Two major questions are: how much does pollution control cost, and how much control are we willing to pay for? Engineers can tell us how much it costs to place abatement devices on existing equipment such as catalytic converters on automobiles or scrubbers on coal fired power plants. But recently people have begun to realize that such remedial solutions may not be the most effective and that pollution can be reduced more economically by re-evaluating fundamental procedures. Thus if gasoline powered automobiles are too polluting, perhaps commuters could use electric cars or more mass transit. Conservation of electricity is a more effective and cheaper way to reduce air pollution from power plants than is the addition of pollution abatement equipment. Advocates of nuclear power claim that the health effects from operating nuclear generators and disposing of their wastes are less than those from burning coal. The information in this chapter will broaden students background for these important discussions.

Chapter 18 contaains an Earth Science and the Environment box on air pollution in Los Angeles.

Answers to Discussion Questions

1. If pure coal is burned completely it produces carbon dioxide and water which are not toxic, but contribute to possible greenhouse warming. However, as explained in the text, coal is never pure but contains impurities such as sulfur and nitrogen compounds and minerals. Thus sulfur oxides, nitrogen oxides, and fly ash are released during complete combustion.

2. No, coal and diesel fuel have different chemical compositions and impurities. In addition, different combustion conditions create different chemical reactions and byproducts.

3. (a) Particulate; (b) gaseous; (c) none; (d) and (e) gaseous and particulate.

4. No. Settling is so slow that a chamber would have to be impossibly large.

5. Oxidation of nitrogen impurities and atmospheric nitrogen to nitric oxide in the cylinder; then further oxidation to nitric acid in the atmosphere.

6. Oxides of nitrogen.

7. Oxidation of sulfur to sulfur dioxide in the flame; further oxidation to sulfuric acid in the atmosphere.

8. Particles fall in air; gases do not. Sulfur dioxide, a gas, does not precipitate until it is adsorbed on dust particles or is oxidized to sulfuric acid which dissolves liquid droplets. The latter process is generally slower and therefore acid rain often occurs further from the source than the deposition of acidic dusts.

9. Soils rich in limestone can neutralize acid rain. Acidic mists are more concentrated in acid than raindrops are, and can do more damage.

10. Warm air above cool is more stable because the warm air is less dense. This stability causes stagnation. It is called an inversion because the temperature in the lower atmosphere normally decreases with altitude.

11. Figure 18-8 shows temperature decreasing continuously with altitude, a condition conducive to rapid dispersal of pollutants. The least favorable conditions occur under an inversion layer, as shown in Figure 18-9.

12. The figure should show temperature increasing with altitude between ground and 200 m, gradually decreasing temperature between 200 m and 1000 m, a second band of rising temperature between 1000 and 1200 m, and then the temperature decreases again with further increases in altitude.

13. When the pollutants finally reach the ground, they have become more diluted, hence, less harmful. However, effects, such as an increase in the carbon dioxide level or acid rain, cannot be prevented by chimneys.

14. Method (a) is uncertain because climates change from a variety of natural causes, and methods (c) and (e) suffer from similar defects. Methods (b) and (d) are much better, because the time spans are so short that climatic drifts are ruled out.

15. Air quality standards do not apply to the concentrations at the stack. Therefore, you would need to know what the emission standards were for the plant in question, to determine whether there was a violation. Alternatively, you could analyze air near the plant at ground level, and then you could use the ambient standards as a guide, but you would have to prove that the plant in question is the source. A final possibility would be to calculate the dilution expected between the stack and the ground level, and apply this factor to the stack concentration to predict the expected ambient ground level concentration. Such calculations are carried out by meteorologists, who use mathematical models that have been tested experimentally.

16. This is a question for class discussion; there are no right or wrong answers.

Selected Bibliography

Douglas G. Brookins: The Indoor Radon Problem. New York, Columbia University Press, 1990. 229 pp.

Russell W. Johnson, Glen E Gordon, William Calkins, and A.Z. Elzerman, eds: The Chemistry of Acid Rain. American Chemical Society Symposium, 1987. 353 pp.

James J. MacKenzie and Mohamed T. El-Ashry: Air Pollution's Toll on Forests and Crops. New Haven, CT, Yale University Press, 1990. 384 pp.

Chris C. Park: Acid Rain: Rhetoric and Reality. London, Metheun Press, 1988. 272 pp.

James L. Regens and Robert W. Rycroft: The Acid Rain Controversy. Pittsburg, University of Pittsburg Press, 1988. 228 pp.

William Vatavuk: Estimating the Costs of Air Pollution Control. Chelsea, MI, Lewis Publishers, 1990. 235 pp.

Chapter 18

Air Pollution

Multiple Choice:

1. When coal burns it produces
(a) carbon trioxide; (b) carbon dioxide; (c) hydrocarbons; (d) oxygen; (e) b and d.

2. The primary source of nitrogen oxide pollution is
(a) volcanic eruptions; (2) lightning; (c) automobile exhaust; (d) coal fired power plants; (e) nuclear power plants.

3. In air pollution terminology, a particle is any pollutant larger than
(a) a molecule; (b) a sand grain; (c) a clay particle; (d) a speck; (e) none of these.

4. Pollutants that are not released directly into the air by any industrial process, but are generated by reactions within the atmosphere are called
(a) primary air pollutants; (b) secondary air pollutants; (c) tertiary air pollutants; (d) fly ash; (e) industrial air pollutants.

5. Fly ash is
(a) released mainly from automobile exhaust; (b) a precursor to acid rain; (c) a precursor to photochemical smog; (d) mineral matter released when coal is burned.

6. The Haagen-Smit experiment was the first to prove that when automobile exhaust is exposed to sunlight it produces
(a) primary air pollutants; (b) soot; (c) fly ash; (d) carbon dioxide; (e) smog.

7. Photochemical smog can be produced artificially by a combination of
(a) automobile exhaust and UV lamps; (b) UV lamps and ozone; (c) gasoline vapor and oxygen; (d) gasoline vapor and UV lamps.

8. Acidic precipitation is produced
(a) when sulfur dioxide and oxides of nitrogen dissolve in water droplets in the atmosphere; (b) whenever fuels are burned; (d) by nuclear fission reactors; (d) by atmospheric inversion; (e) when sunlight reacts with automobile exhaust.

9. The major source of sulfur dioxide pollution is

(a) automobile exhaust; (b) coal-fired electric generators; (c) home furnaces; (d) evaporation of paints and pesticides; (e) nuclear power plants.

10. Under normal conditions, as the Sun warms the Earth and its lower atmosphere, warm air rises to mix with cooler air and (a) air pollutants are concentrated close to the ground; (b) air pollutants are diluted near the ground; (c) turbulence is created; (d) smog is produced; (e) b and c.

11. Atmospheric inversions aggravate the effects of air pollutants because
(a) they tend to occur during bright sunlight when photochemical effects are produced; (b) they help to make the pollutants more concentrated; (c) they trap pollutants and slow down their biodegradation; (d) toxic effects are more serious when the air is stagnant.

12. During an inversion
(a) air at higher elevations is cooler than air near the ground; (b) air at higher elevations is warmer than air near the ground; (c)the air temperature is the same near the ground as it is at higher elevations; (d) pollutants are readily dispersed.

13. An example of an epidemiological study is:
(a) Health records of workers in a chemical plant are compared with those of farmers living in the surrounding countryside. (b) Rats are fed high doses of a food additive and tested to see how many contract cancer. (c) Air pollutants are mixed in an experimental room which is lit up by sunlamps. (d) Atmospheric temperature is recorded as a function of altitude and the temperature profile is compared with the concentration of pollutants.

14. In the United States the cost of deterioration of buildings and materials from air pollution is
(a) a minor problem; (b) estimated at several billion dollars per year; (c) not a problem yet; (d) a problem that will be easy to solve.

15. Measures that conserve heat by reducing air leaks usually (a) increase indoor pollution; (b) decrease indoor pollution; (c) prevent indoor pollution; (d) cause indoor pollution; (e) are not cost efficient.

16. ______ is directed to monitor the purity of the air in the United States
(a) congress; (b) the local police department; (c) the EPA; (d) the president; (e) a group of scientists

17. Nuclear warfare
(a) would have no affect on the atmosphere; (b) would be likely to affect our climate adversely; (c) could create a nuclear winter; (d) would lift huge amounts of pulverized soil into the stratosphere; (e) b, c, and d.

True or False:

1. Twenty people died from the effects of air pollution in Donora, Pennsylvania in 1948.

2. Benzene, a product of incomplete combustion of fossil fuels, is a carcinogen.

3. Volatiles pose no threat to our health.

4. A pesticide called toxaphene which is used in the cotton fields of the South has been found in the mud of a lake in Northern Minnesota.

5. Smog causes no damage to vegetable plants.

6. Ozone is both a pollutant and a beneficial component of the atmosphere, depending on elevation.

7. A rain storm in Baltimore was as acidic as vinegar.

8. An atmospheric inversion can persist for a couple of hours only.

9. Laboratory studies are always conclusive.

10. When the Geneva Steel mill shut down, the concentration of particulates in the air decreased, and the number of children hospitalized for bronchitis, asthma, pneumonia, and pleurisy declined by 35 percent.

11. Air pollution leads to an estimated $3 to $4 billion in crop damage every year in the United States.

12. Indoor air is generally more polluted than outdoor air.

13. Even if each factory, power plant, and automobile complies with the Clean Air Act, if the total pollution level exceeds National Ambient Air Quality Standards, then the EPA must set stricter controls.

14. In general, pollution control becomes more expensive as higher proportions of pollutants are removed.

15. Fires started during the Gulf War will seriously affect global climate.

Completion:

1. Today the primary global source of sulfur dioxide pollution is ______ ______ ______.

2. ______ are compounds composed of carbon and hydrogen.

3. A/an ______ compound is one that evaporates readily and therefore easily escapes into the atmosphere.

4. Gases and particulates released during combustion and manufacturing are called ______ ______ ______.

5. Smog is produced when automobile exhaust is exposed to ______.

6. When coal burns, some of minerals escape from the chimney as ______ ______ which settles out as gritty dust.

7. ______ in the stratosphere absorbs ultraviolet radiation.

8. ______ ______ reacts in moist air to produce sulfuric acid.

9. When the ground and air close to the ground cool at night a condition called ______ ______ may occur.

10. ______ is the study of the distribution and determination of health and its disorders.

11. Tree death in West Germany has been attributed to ______ ______.

12. A radioactive gas that is released from soil, stone, and building materials is ______.

13. The major legislation that dictates Federal air Pollution Policy is the ______ ______ ______.

14. ______ ______ balances the monetary cost of pollution control against the monetary benefits of having the pollutants removed.

15. The atmospheric condition that might result from a nuclear war has been named ______ ______.

Answers for Chapter 18

Multiple Choice: 1. b; 2. c; 3. a; 4. b; 5. d; 6. e; 7. a; 8. a; 9. b; 10. e; 11. b; 12. b; 13. a; 14. b; 15. a; 16. c; 17. e

True or False: 1. T; 2. T; 3. F; 4. T; 5. F; 6. T; 7. T; 8. F; 9. F; 10. T; 11. T; 12. T; 13. T; 14. T; 15. F

Completion: 1. coal-fired electric generators; 2. hydrocarbons; 3. volatile; 4. primary air pollutants; 5. sunlight; 6. fly ash; 7. ozone; 8. sulfur dioxide; 9. atmospheric inversion; 10. epidemiology; 11. acid precipitation; 12. radon; 13. clean air act; 14. cost-benefit analysis; 15. nuclear winter

CHAPTER 19

Motion in the Heavens

Discussion

Astronomers study objects so far away that they can never sample them directly. During the study of cosmology, they infer events that occurred billions of years before the formation of the Earth, at the very instant when space and time were created. Thus our knowledge of the heavens cannot be derived by direct sampling. Rather it is accumulated by interpretation of data to build inferences about events and objects outside the realm of direct experience.

The first great astronomical problem was to explain the motion of the Earth, Sun, planets, and stars --a task so difficult and so compelling that it became a central issue of scientific debate from Aristotle to Newton. In fact, the drive to understand motion in the heavens forced people to alter their patterns of reasoning and became the foundation of modern science.

In this chapter we explain Aristotle's geocentric Universe and trace the evolution of thought to Galileo's heliocentric Solar System. We then explain how Galileo's conclusions were supported by Newton's conclusion about the universality of gravity. With this background, we review the modern understanding of the motions in the Solar System -- planetary motion, the motion of the Moon, and eclipses. Finally, as an introduction to the two following chapters, we introduce the techniques of modern astronomy.

Chapter 19 conatins two Focus On boxes: The Constellations and What Orbits What?.

Answers to Discussion Questions

1. (a) rotation on its axis; (b) revolution around the Sun; (c) rotation on its axis.

2. (a) the Earth's rotation; (b) both the Earth's rotation and the Moon's revolution; (c) the Moon's revolution.

3. Aristotle's conclusions were based both on observation and dogma. No modern editor would accept a theory based on the belief that celestial spheres are an expression of God's will. However, Aristotle also reasoned that if the Earth revolved around the Sun, then he should observe a stellar parallax shift. Since such a shift was not observed, he reasoned that the Earth

must be stationary. Such reasoning was perfectly logical given the data. Therefore if you were a journal editor in ancient Greece, this line of reasoning would be valid.

4. Older style speedometers consist of a flat plate with numbers printed on it and a moving needle. In most cars the needle is positioned above the number plate. Let us say that the car is moving at 80 km/hr. The driver, looking directly at the instrument, sees the needle in line with the 80 km/hr reading. But the passenger sees the instrument at an angle and according to his line of sight, the needle appears to lie over the 70 km/hr mark. This misreading is the parallax error. Point out to the student that on scientific instruments, there is often a mirror behind the indicator needle. If the observer lines up the needle and the mirror, no parallax error can occur.

5. It would be easier to estimate the distance to nearby stars. The parallax angle is greatest when the observer moves an appreciable distance compared to the distance to the object being studied. Since the diameter of the Earth's orbit is small compared to the distance to even the nearest stars, the angles measured are a very small fraction of a degree. The angles become so small for distant objects that they become impossible to measure.

6. There would be an eclipse of the Sun every new Moon and an eclipse of the Moon every full moon.

7. There are approximately 29 1/2 days in a lunar month and therefore a 12 month lunar year is shorter than a solar year. In some medieval European societies the time difference between the two calendars was considered to be a blank time, and these days were set aside as a winter festival.

Selected Bibliography

Refer to two excellent standard texts.

George O. Abell, David Morrison, and Sidney Wolff: Exploration of the Universe 6th ed. Philadephia, Saunders College Publishing, 1991. 682 pp.

Jay M. Pasachoff: Journey Through the Universe. Philadephia, Saunders College Publishing, 1992. 391 pp.

Chapter 19

Motion in the Heavens

Multiple Choice:

1. If we knew that the Sun rose every morning in the east and set every evening in the west, approximately 12 hours later, but we had no other information, which of the following conclusions would be reasonable?
(a) The Sun orbits around the Earth once every 24 hours. (b) The Sun orbits around the Earth once every 365 days. (c) The Earth orbits around the Sun once every 365 days. (d) The Earth rotates on its axis once every 365 days.

2. In the Northern Hemisphere, the Pole Star
(a) remains motionless and all other stars revolve around it; (b) moves around all the other stars; (c) disappears completely every 29.5 days; (d) spills water at 2 o'clock in the morning.

3. The model that we use to describe the motion of the heavenly bodies
(a) is the geocentric model; (b) the heliocentric model; (c) is the same one proposed by ancient Greek philosophers; (d) was first proposed by Copernicus; (e) b and d.

4. Kepler calculated that the planets move in
(a) circular orbits; (b) angular orbits; (c) elliptical orbits; (d) retrograde motion; (e) parallax.

5. The reason that planets move in curved lines rather than flying off into space was first explained by
(a) Copernicus; (b) Galileo; (c) Aristotle; (d) Newton; (e) Brahe

6. The Earth spins approximately ______ times for each complete orbit around the Sun.
(a) 29.5; (b) 2; (c) 365; (d) 12; (e) 100.

7. Different stars and constellations are visible during different seasons because
(a) The Sun's revolution around the Earth changes our view of the night sky; (b) The Earth's revolution around the Sun changes our view of the night sky; (c) parallax; (d) the Earth is the center of the Universe; (e) the Earth is so large.

8. The constellation Orion is visible from North America during winter but is not seen on summer evenings. Which of the following statements is true?
(a) The stars in Orion pulsate so that they shine brightly for six months and then are dark for six months. (b) Orion is visible

from South America during the summer. (c) Orion is visible from the opposite side of the Earth, India, during the summer. (d) Orion cannot be seen during the summer because the Sun is located between Orion and the Earth.

9. The phase of the Moon depends on the relative positions of (a) the Sun, Moon, and Earth; (b) the Moon and Earth; (c) the Sun and Earth; (d) all of the planets; (e) the Moon and Sun.

10. When the Earth lies directly between the Sun and Moon it produces a/an
(a) solar eclipse; (b) absorption spectrum; (c) partial eclipse; (d) constellation; (e) lunar eclipse.

11. which of the following statements about eclipses is true? (a) A solar eclipse occurs every full moon. (b) A lunar eclipse occurs every full moon. (c) A solar eclipse occurs every new moon. (d) A lunar eclipse occurs every new moon. (e) none of these.

12. As light passes from the hot interior of a star through the cooler outer layers,
(a) some wavelengths are selectively absorbed in the star's outer atmosphere; (b) some wavelengths are selectively reflected in the star's outer atmosphere; (c) some of the star melts; (d) a rainbow is produced; (e) hydrogen fusion is initiated.

13. Absorption spectra can be used
(a) to determine the chemical composition of a star; (b) to determine surface temperature and pressures of star; (c) a and b; (d) to weigh a star.

True or False:

1. Constellations appear and disappear with the seasons.

2. Planets sometimes reverse direction and drift westward.

3. Ptolemy modified the geocentric model to explain the retrograde motion of the planets.

4. Galileo proposed the geocentric universe.

5. The geocentric universe easily explains retrograde motion.

6. Mars has to reverse direction to catch up with Earth.

7. A new Moon occurs when the Moon is on the opposite side of the Earth from the Sun.

8. A planet moves in a straight line unless a force is exerted on it.

9. As the Earth rotates, its axis wobbles.

10. We always see the same lunar surface, and the other side was invisible to us until the Space Age.

11. Normally the Moon lies out of the plane of the Earth's orbit around the Sun.

12. On each successive evening the Moon rises about 53 minutes earlier.

13. The Moon does not emit its own light but reflects light from the Sun.

14. More than 99.99 percent of the spectrum is invisible to the naked eye.

15. Astronomers discovered the element helium in the Sun 27 years before it was discovered on Earth.

Completion:

1. Groups of stars are called ______.

2. ______ change position with respect to the stars.

3. Aristotle proposed a ______ Universe.

4. In the geocentric Universe, a series of concentric ______ ______ made of transparent crystal surround the Earth.

5. ______ is the apparent change in position of an object due to the change in position of the observer.

6. Kepler calculated that the planets moved in ______ orbits.

7. ______ proposed the heliocentric universe.

8. A person does not fall off the Earth as it moves because of ______.

9. ______ was the first person to see that the Moon had mountains, hills, craters, and large plains.

10. ______ was the first person to record dark spots on the Sun.

11. ______ ______ between the Sun and a planet forces the planet to move in an elliptical orbit.

12. The Moon's ______ causes tides on Earth.

13. When the Moon passes directly between the Earth and the Sun a/an ______ ______ occurs.

14. The shadow band where the Sun is totally eclipsed is called the ______.

15. Telescopes that collect light with large curved mirrors are called ______ telescopes.

Answers for Chapter 19

Multiple Choice: 1. a; 2. a; 3. e; 4. c; 5. d; 6. c; 7. b; 8. d; 9. a; 10. e; 11. e; 12. a; 13. c

True or False: 1. T; 2. F; 3. T; 4. F; 5. F; 6. F; 7. F; 8. T; 9. T; 10. T; 11. T; 12. F; 13. T; 14. T; 15. T

Completion: 1. constellations; 2. planets; 3. geocentric; 4. celestial spheres; 5. parallax; 6. elliptical; 7. Copernicus; 8. inertia; 9. Galileo; 10. Galileo; 11. gravitational attraction; 12. gravitation; 13. solar eclipse; 14. umbra; 15. reflecting

CHAPTER 20

Planets and Their Moons

Discussion

The space age began on October 4, 1957 when the U.S.S.R launched Sputnik, the first spacecraft. Most of what we know about the planets except for their motions, sizes, and masses, has been learned in the past 35 years, one half a lifetime.

Of the millions of striking images recorded by dozens of spacecraft, one of the most memorable was taken by Voyager II in 1989 as it sped past Neptune and on toward interstellar space. With its mission virtually completed, cameras were aimed back toward the inner Solar System and Voyager transmitted 60 photographs of the solar system as seen from its outer fringe. The Earth and its planetary neighbors appeared as small pinpoints of light in the blackness of space. Voyager was one of the great achievements of the first generation of spacecraft to explore our Solar System. In a speech commemorating this age of discovery, Carl Sagan said:

> "There is no evidence -- not a smidgen of evidence -- suggesting life anywhere else. And for me, that underscores the rarity and preciousness of the Earth and the life upon it."

One important point of our study of planetary geology is that all the protoplanets were composed of the same elements in roughly the same proportions. The major differences were the initial masses and distances from the Sun. Today the planets ar compositionally quite different from one another, and only one, our Earth, supports life. Thus, initial differences in mass and distance from the Sun caused chemical and physical changes that amplified to create different environments on each planets and the Moon.

We introduced the evolution of the Universe and Solar Syste in Chapter 1 to establish a broad palette for our understanding of the formation and structure of the Earth. The origin of the Solar System is re-introduced here, in Chapter 20, but some professors may wish to review the earlier material. Chapter 20 contains an Earth Science and the Environment box, Impacts of Asteroids with the Earth and a Focus On box entitled Extraterrestrial Life.

Answers to Discussion Questions

1. Lower (and night-time temperatures would be warmer). If the sunny side received sunlight for a shorter period of time, there would be less time for rock and soil to heat up.

2. We know that it would have taken roughly 800 million years of radioactive decay to build up enough heat to melt the rocks on the Moon. If the oldest igneous rock on the Moon were formed when the Moon was 800 million years old, we wouldn't know whether the Moon was heated solely by radioactive decay or by some combination of processes. For example, a period of intense meteor bombardment could have occurred early in the Moon's history but all the igneous rock thus formed was remelted when the Moon was reheated by radioactive decay. Alternatively, about 800 million years after the Moon's formation, intense external bombardment could have coincided with a heat buildup from radioactive decay to melt the lunar rock. In the absence of additional data we could not distinguish between these two possibilities.

3. The Earth is geologically much more active than the Moon, and old rocks are continuously being forced into the mantle to be reheated and reformed. Also, weathering and erosion alter surface features. On the Moon, ancient rock and landforms have been preserved, leaving clear traces of ancient lava flows and periods of intense meteor bombardment. Since the Moon and the Earth occupy the same region of space, a meteor bombardment that occurred on the Moon must also have affected the Earth. Therefore, by studying the effects of the meteor bombardment on the Moon, we can deduce that the same events occurred on the Earth.

4. No, probably not. For conditions on Venus to be comparable to those on Earth, the carbon dioxide concentration in the atmosphere would have to be reduced to a percent or less. The temperature on that planet is so high now that a temperature drop of 20° would not shift the equilibrium enough to cause widespread chemical changes of gaseous carbon dioxide to chemically bound and dissolved forms.

5. The surface of Venus is so much hotter than that on Earth that the lithosphere may be more plastic. Thus a mantle plume may distort and fracture surface rocks, but be unable to form lithosphere-deep cracks. Alternatively, since Venus is smaller than Earth, its mantle is cooler and it has a thicker lithosphere, which might not fracture as easily as the Earth's lithosphere does.

6. Venus exhibits tectonic behavior but no horizontal plate motion as explained in question 5, above. In addition, the greenhouse warming of Venus, explained in the text, has led to dramatically different atmospheric and climatic conditions.

7. Figure 20-9 shows two types of surfaces, a lowland plain wit impact craters and a mountain range. The mountains were formed by tectonic activity after the period of extensive impact bombardment. The older craters have not been obliterated by erosion, indicating that wind, ice, and running water are not active agents on the Martian surface. Figure 20-14 shows dry river canyons pockmarked by impact craters. Thus water must hav once been present on the surface. The impact craters overlie some of the stream beds. Since the last active period of meteorite bombardment in the inner Solar System was about 3.9 billion years ago, we know that surface water vanished from Mars by that time.

8. (a) We could deduce that the planet was probably quite cold internally and that no significant volcanic or tectonic activity had occurred for quite a long time. We would also know that weathering and erosion had occurred, so we could conclude that the planet had or has liquids on its surface and an atmosphere. (b) These gases are commonly expelled by volcanic action. Therefore, we could conclude that the planet has recently experienced intense volcanic activity. (c) We can conclude that there were two periods of meteor bombardment interspersed with periods of tectonic or volcanic activity. The geological histor can be reconstructed as follows: After the initial formation of the planet, the entire surface was subject to meteor bombardment. Later, large areas were altered by geological activity. Then a second, smaller, meteor bombardment occurred, followed by a second era of tectonic activity that affected only a part of the planet. The topography of the three regions can by explained by this general scheme. The region covered by many impact craters is the oldest. The large craters were formed during the initia period of meteor bombardment. The smaller craters, some of whic lie inside the larger ones, were formed during the second period of meteor bombardment. The third of the planet that is smoother and scattered with a few small craters experienced tectonic change after the first meteor bombardment. Since any leveling b wind or water would have eroded craters across the entire planet, we can infer that global erosion did not occur. Instead a smoot lava flow covered the large craters in this region. This lava flow may have been similar to the ones that formed the marias on the Moon. The second period of meteor bombardment followed this volcanic activity. The mountainous region of the planet is geologically the youngest. Since there are no craters visible here, the surface must have been shaped after the last period of meteor bombardment. The absence of canyons or river basins

supports the belief that there is and has been a relatively diffuse atmosphere on this planet.

9. Jupiter is the most star-like planet, being composed mainly of hydrogen and helium. Furthermore, Jupiter continues to contract, and produces energy from its gravitational coalescence. If it were more massive the contraction would eventually heat its interior enough to initiate fusion and Jupiter would become a star. This scenario is supported but the observation that many pairs of stars exist in our galaxy. Thus, Arthur C. Clarke's fiction is based on credible science. However, calculations show that Jupiter is not massive enough to undergo the transformation.

10. If the Sun grows appreciably hotter, some of the hydrogen and helium on Jupiter will be boiled off into space and the planet will become less massive. It will acquire a denser, more solid form, with a greater percentage of heavy elements.

11. The rings are primarily made of dust and rock. Each particle has its own orbit. The mass of a gas molecule is so much less than that of even the tiniest piece of dust that most of the gases would have escaped off into space.

Selected Bibliography

Several books on planetary astronomy are:

J. Kelly Beatty and Andrew Chaikin: The New Solar System 3rd ed. Cambridge, Cambridge University Press, 1990. 326 pp.

William David Compton: Where No Man Has Gone Before: A History of Apollo Lunar Exploration Missions. Washington, DC, USGPO, 1990. 415 pp.

Mark Littmann: Planets Beyond: Discovering the Outer Solar System. New York, Wiley and Sons, 1988. 319 pp.

Bruce Murray: Journey into space: The First Thirty Years of Space Exploration. New York, W.W. Norton, 1989. 381 pp.

Horton E. Newsom and John H. Jones, et al eds: Origin of the Earth. Oxford, Oxford University Press, 1990. 378 pp.

Chapter 20

Planets and Their Moons

Multiple Choice:

1. The Solar System was originally
(a) a diffuse collection of hot fusing hydrogen and helium; (b) a cold dense cloud of dust and gas; (c) a cold diffuse cloud of dust and gas; (d) composed solely of hydrogen and helium.

2. The protosun originally grew hot because
(a) the primordial dust was already hot; (b) gravitational collapse of its original cloud heated the dust and gas; (c) hydrogen fused into helium; (d) it was energized by a black hole; (e) hydrogen atoms reacted chemically with helium, oxygen, and other elements.

3. The modern Sun was born when
(a) pressure inside the protosun became great enough for fusion to begin; (b) pressure inside the protosun became great enough for fusion to cease; (c) it was energized by a black hole; (d) hydrogen atoms reacted chemically with helium, oxygen, and other elements.

4. The fact that Mercury has retained craters from meteorite bombardment that occurred during the early history of the Solar System while the Earth has not, tells us
(a) that the Earth is a different type of planet; (b) that the Earth had a protective shield to protect it from impact; (c) that tectonic activity and erosion have occurred on Mercury during the last four billion years; (d) that tectonic activity and erosion have not occurred on Mercury during the last four billion years.

5. The atmosphere of Venus is so different from the atmosphere of Earth because
(a) Venus is closer to the Sun than Earth is; (b) Earth is closer to the Sun than Venus is; (c) the atmosphere of Venus has less carbon dioxide than Earth has; (d) a and c.

6. The surface of the Moon is covered with
(a) igneous rocks; (b) metamorphic rocks; (c) sedimentary rocks; (d) all three types of rocks; (e) half rock and half water.

7. Which planet has almost no atmosphere, a cratered surface, and is extremely hot during the day and extremely cold at night?
(a) Mercury; (b) Venus; (c) Earth; (d) Mars; (e) Jupiter; (f) Pluto.

8. Which planet has a dense atmosphere, rocky surface, and surface temperatures of about 450°C?
(a) Mercury; (b) Venus; (c) Earth; (d) Mars; (e) Jupiter; (f) Saturn; (g) Pluto.

9. The Moon formed
(a) shortly before the Earth; (b) about 4.6 million years ago; (c) shortly after the Earth, about 4.6 billion years ago; (d) about 40 billion years ago.

10. Why has the Moon remained geologically quiet and inactive compared to the Earth for the past 3 billion years?
(a) The Moon is so much larger that it took longer to cool. (b) The Moon is so much smaller that it soon cooled. (c) The Moon's core is solid. (d) The lunar crust remained liquid. (e) none of these.

11. Which of the following statements is not correct about Mars?
(a) Liquid water used to flow across the Martian surface but now it is dry. (b) The largest volcano in the Solar systems is found on Mars. (c) Ice caps on Mars are mainly frozen carbon dioxide. (d) The surface of Mars contains cratered regions and younger lava plains. (e) The Martian atmosphere is dense with a high concentration of carbon dioxide.

12. The chemical composition of Jupiter is most similar to that of
(a) Mars; (b) the Earth; (c) the Moon; (d) Io; (e) Titan; (f) the Sun.

13. Which moon has a dense methane atmosphere, may have methane lakes and icecaps, and a mean surface temperature of about -180°C ?
(a) Io; (b) Europa; (c) Ganymede; (d) Titan.

14. Above the surface of Jupiter the atmosphere consists of
(a) liquid metallic hydrogen; (b) liquid hydrogen; (c) oxygen, nitrogen, carbon dioxide, argon, and water; (d) hydrogen, helium, ammonia, methane, water, hydrogen sulfide, and other compounds.

15. It is now believed that the interior of Io has been heated primarily by
(a) the Sun; (b) the hot, radioactive core of Jupiter; (c) energetic, electrically charged particles that are accelerated by Jupiter's strong magnetic field; (d) gravitational forces from Jupiter, Europa, Ganymede, and Callisto; (e) meteor bombardment.

16. The rings of Saturn are believed to be
(a) an auroral display of ionized gases; (b) particles of pure ice; (c) the remains of a hurricane-like storm that was first

recorded by Galileo; (d) a solid ring like a phonograph record; (e) small particles of dust, rock, and ice.

17. Which of the following statements about Uranus is not correct?
(a) Uranus is much larger than the Earth. (b) Uranus is much denser than the Earth. (c) Uranus has rings. (d) Uranus is composed of hydrogen and helium with a mineral core.

18. Which of the following orders of sizes is correct?
(a) Earth > Jupiter > the Moon; (b) Jupiter > the Moon > Earth; (c) Pluto > Saturn > the Moon; (d) Saturn > Mars > Earth; (e) Uranus > Earth > Mercury.

19. The tens of thousands of smaller objects that orbit the Sun between Mars and Jupiter are called
(a) meteoroids; (b) asteroids; (c) comets; (d) meteors; (e) none of these.

20. A ______ is composed of ice mixed with bits of silicate rock, metals, and frozen crystals of methane, ammonia, carbon dioxide, carbon monoxide, and other compounds.
(a) meteoroid; (b) chondrule; (c) comet; (d) meteor; (e) none of these

True or False:

1. The outer Jovian planets, Jupiter, Saturn, Uranus, and Neptune lost most of their hydrogen, helium, and other light elements.

2. Mercury orbits faster than any other planet.

3. Tectonic activity has ceased on Mercury.

4. The atmosphere of Venus is 90 times more dense than the Earth's atmosphere.

5. Venus has mountains higher than Mount Everest.

6. The Moon's maria are large oceans of hydrogen.

7. The Moon has a metallic core surrounded by silicate rocks of lower density.

8. Because the lithosphere on Mars is stationary, volcanic activity can continue in one place.

9. The annual vegetation blooms people thought that they saw on Mars are actually huge dust storms.

10. Jupiter, Saturn, Uranus, Neptune, and Pluto are all gaseous giants, composed mainly of hydrogen and helium.

11. Io does not have active volcanoes.

12. Volcanoes are now frequent on Europa.

13. Saturn would float on water if there were a basin large enough to hold it.

14. Titan's large size and low temperature allow it to have an atmosphere.

15. Pluto has never been visited by spacecraft.

16. Asteroids change their orbits frequently and erratically.

17. Organic molecules have been detected in dust clouds deep in interstellar space.

Completion:

1. When pressure inside the protosun then became great enough for to ______ begin, and the modern Sun was born.

2. Mercury, Venus, Earth, and Mars are mostly solid spheres and are called the ______ ______.

3. The temperature on the sunny side of ______ reaches 450°C, while the shady side is frigidly cold.

4. Venus is much hotter than Earth because of a runaway ______ ______.

5. The lunar maria formed when ______ filled circular meteorite craters.

6. Geologists think that a rising ______ ______ may have formed the Tharsis bulge on Mars.

7. ______ planets have dense gaseous atmospheres, very large liquid interiors, and much smaller solid cores.

8. Jupiter's middle layer, between its core and outer sea of liquid hydrogen, is composed of ______ ______ ______.

9. The ______ ______ ______ is a giant hurricane-like storm that has remained on the surface of Jupiter for many years.

10. The rings of Saturn may be the debris of one or more ______ that got too close to the planet.

11. Titan is the only moon in the Solar System with an appreciably ______ atmosphere.

12. Europa is a moon of ______.

13. _____ is the smallest planet in the Solar System.

14. A fallen meteoroid is called a/an ______.

15. Most stony meteorites contain small, round grains about 1 millimeter in diameter called ______, that are composed largely of olivine and pyroxene.

Answers for Chapter 20

Multiple Choice: 1. c; 2. b; 3. a; 4. d; 5. a; 6. a; 7. a; 8. b; 9. c; 10. b; 11. e; 12. f; 13. d; 14. d; 15. d; 16. e; 17. b; 18. e; 19. b; 20. c

True or False: 1. F; 2. T; 3. T; 4. T; 5. T; 6. F; 7. T; 8. T; 9. T; 10. F; 11. F; 12. F; 13. T; 14. T; 15. T; 16. T; 17. T

Completion: 1. fusion; 2. terrestrial planets; 3. Mercury; 4. greenhouse effect; 5. lava; 6. mantle plume; 7. Jovian; 8. liquid metallic hydrogen; 9. great red spot; 10. moons; 11. dense; 12. Jupiter; 13. Pluto; 14. meteorite; 15. chondrules

CHAPTER 21

Stars, Space, and Galaxies

Discussion

In this chapter we introduce the structure of our Sun, followed by the life and death of stars. Dying stars produce white dwarfs, neutron stars, or black holes depending on their mass.

Finally, we discuss galaxies, quasars, and the evolution of the Universe. A brief review of the big bang theory introduced in Chapter 1 would be appropriate here. We were fortunate that data from the cosmic background explorer spacecraft (COBE) was released in the spring of 1992, as we were reading the fianl proofs. This information showed that the Universe was non-homogeneous in its earliest infancy. Thus one major objection to the big bang theory has been removed. Perhaps astrophysicists will soon solve the mystery of dark matter and consolidate our understanding of the Universe.

Answers to Discussion Questions

1. No, sound travels only through matter and cannot travel through the vacuum of space as electromagnetic radiation does.

2. Yes, all waves exhibit Doppler shifts. If you were in front of a moving boat you would observe more waves per second than if you stood behind it.

3. No, Hubble's law applies to galaxies, not stars within the Milky Way. All the galaxies are believed to be flying away from each other, but within a given galaxy stars orbit around the nucleus.

4. The Moon and the planets don't emit light but reflect sunlight; we learn little from studying this reflected light.

5. No, in order to study a distant object as it exists today, information would have to travel faster than the speed of light, which is theoretically impossible.

6. No. The temperature in the Sun is so high that molecules such as H_2, O_2, and H_2O cannot exist. Also, there is little elemental oxygen in the Sun, and finally, ordinary chemical reactions release such small quantities of energy compared with

nuclear fusion reactions that their effect would be insignificant.

7. The density of the Sun is regulated by a dynamic balance between the force of gravitation pulling inward and that of radiation pushing outward. The gravitational force diminishes with distance from the core but the radiation remains strong, so the gases are diffuse near the surface of the Sun.

8. You would know that the star had passed through its hydrogen fusion stage, had probably expanded to become a red giant, contracted again until helium started to fuse, and now the helium fuel was largely consumed. The star would soon experience further drastic changes but the nature of these changes would depend on its mass.

9. A white dwarf is an old, burned out, dense star no longer undergoing fusion. A red giant is still providing energy by fusion. Yes, a star about the mass of the Sun will be a red giant after hydrogen fusion slows down, and will become a white dwarf after the helium is consumed.

10. The star and the planets in a Solar System are initially formed from a single cloud of dust and gas. If that cloud originally contained no heavy elements, any planets would be composed solely of hydrogen and helium. We would not expect life on those planets because life depends on the existence of heavier elements.

11. They are older than our Sun. Our Sun has more heavy elements that are believed to have been formed from the remnants of supernova explosions. Stars with fewer heavy elements originated during an earlier period of galactic evolution.

12. (a) Stars are too large to emit such a sharp radio pulse, and they cannot possibly rotate so fast. (b) A planet orbits a star so its position relative to the stars changes continuously. Also, planets do not emit such large quantities of energy. (c) A galaxy is much too large to emit such a sharp radio pulse and cannot possibly rotate so fast. (d) Magnetic storms do not oscillate in a regular, unvarying, periodic manner.

13. If a black hole were a permanent resident in our Solar System, it would distort the orbits of planets and comets and otherwise perturb the system severely enough to be detected.

14. The best evidence indicates that few black holes exist within the Milky Way's galactic disk and there is a slim chance that a rocket ship would pass into the field of a black hole. Black holes adjacent to stars cause X-ray emissions that provide

warning. If a rocket did pass close to one, the craft would be pulled off course by the intense gravitational field. If the navigator was alert enough to notice the problem before the ship fell far into the field, and the rocket engines were sufficiently powerful, the crew could escape unharmed. But it would be a real thriller. We know so little about intergalactic space that it is hard to say -- some theories predict that dark matter is composed of intergalactic black holes.

15. A quasar is much larger than a star, much smaller than a galaxy, and much more energetic than either. As explained in the text, it may be a galaxy in formation.

16. a < e < d < c < f < b.

17. Planet-like objects and black holes do not radiate energy. A few years ago, astrophysicists suggested neutrinos might be candidates, but experiments have failed to verify this theory. Scientists have recently suggested that dark matter is made up of a type of particle that we have not yet detected.

Selected Bibliography

Stephen J Hawking: A Brief History of Time From the Big Bang to Black Holes. Toronto, Bantam Books, 1988. 198 pp.

Edward W. Kolb and Michael S. Turner: The Early Universe. Redwood City, CA, Addison-Wesley, 1990. 547 pp.

Lawrence M. Krauss: The Fifth Essence: The Search For Dark Matter. New York, Basic Books, 1989. 230 pp.

Igor Novikov: Black Holes and the Universe. Cambridge, Cambridge University Press, 1990. 176 pp.

Michael Riordan and David N. Schramm: The Shadows of Creation: Dark Matter and the Structure of the Universe. San Francisco, W.H. Freeman, 1991. 278 pp.

Donat G. Wentzel: The Restless Sun. Washington, DC, Smithsonian Press, 1989. 279 pp.

Chapter 21

Stars, Space, and Galaxies

Multiple Choice:

1. One light year is
(a) the distance traveled by light in a year; (b) 9.5 trillion kilometers; (c) 10,000,000 kilometers; (d) 1/3600 of a degree; (e) a and b; (f) a and c.

2. Different regions of the Sun have different temperatures. Which sequence of temperatures is correct, starting with the hottest region first?
(a) core > photosphere > corona; (b) core > corona > photosphere; (c) photosphere > core > corona; (d) corona > core > photosphere.

3. Sunspots are areas
(a) where the temperature is 1000° cooler than surrounding areas; (b) that appear as dark spots on the Sun; (c) caused by the Sun's magnetic field restricting solar turbulence; (d) that appear regularly on the Sun's surface; (e) all of these.

4. The Sun looks so big and bright because it is
(a) the largest star; (b) the closest star; (c) the brightest star; (d) composed mainly of hydrogen and helium; (e) the most energetic star.

5. All main-sequence stars are composed primarily of
(a) hydrogen and helium; (b) hydrogen and oxygen; (c) helium and cesium; (d) hydrogen; (e) helium.

6. The major reason for differences in temperature and luminosities among main-sequence stars is that
(a) some have more hydrogen than others; (b) some are more massive than others; (c) some are younger than others; (d) some are older than others.

7. The size and density of a star is determined by
(a) gravity; (b) fusion; (c) the balance between gravity and fusion; (d) the number of sunspots; (e) none of these.

8. When hydrogen fusion ends within the core of a star, the core initially
(a) expands and becomes hotter; (b) contracts and becomes cooler; (c) expands and becomes cooler; (d) contracts and becomes hotter.

9. When most of the hydrogen in a star's core has been converted to helium

(a) it is in its youngest stage; (b) gravitational contraction cools the core; (c) hydrogen fusion ends in the core; (d) the core becomes a black hole; (e) hydrogen fusion ends in the outer shell.

10. A red giant
(a) is in the youngest stage of a star; (b) has an extremely hot surface; (c) has an inner core of fusing hydrogen; (d) has an inner core of fusing helium; (e) has an outer shell of fusing hydrogen.

11. A star's ______ determines whether it becomes a white dwarf or a supernova.
(a) shape; (b) initial density; (c) initial mass; (d) color; (e) age

12. Population I stars begin life
(a) with heavy elements inherited from population II stars; (b) with only light elements such as hydrogen and helium; (c) before population II stars; (d) during solar eclipses.

13. Neutron stars
(a) are dense stars where neutrons are fusing together to form iron; (b) are composed of approximately 75% hydrogen, 24% helium, and 1% other elements; (c) are compressed so tightly that the electrons and protons in the star are squeezed together to form neutrons; (d) are so dense that light cannot escape from them.

14. Black holes are difficult to detect because
(a) they all lie within the cores of massive galaxies and are obscured by clouds of interstellar dust; (b) visible and radio frequency signals emitted from them are weak and erratic; (c) they emit or reflect no light whatsoever; (d) scientists believe that there are only three or four of them in the Universe.

15. What is the diameter of the disk of the Milky Way galaxy?
(a) 100,000,000 km; (b) 10 light years; (c) 1,000 light years; (d) 100,000 light years; (e) 10 billion light years.

16. Which statement is <u>not</u> true about the Milky Way galaxy.
(a) The nucleus of the Milky Way may contain a massive black hole. (b) Within the past 100,000 years, a giant explosion occurred near the center of the Milky Way. (c) The nucleus of the Milky Way is obscured by dust and gas. (d) The concentration of stars near the galactic nucleus is one million times less than in the outer disk.

17. Scientists can determine the velocity of a distant star by studying the Doppler shift of the light received. The scientists

are actually studying a small change in the __________ of the observed light.
(a) frequency; (b) amplitude; (c) velocity; (d) intensity

18. Hubble's Law states that
(a) the most distant galaxies are moving toward us at the greatest speeds; (b) the most distant galaxies are moving away from us at the greatest speeds; (c) the closest galaxies are moving toward us at the greatest speeds; (d) the closest galaxies are moving away from us at the greatest speeds.

19. A quasar
(a) releases about as much energy as a normal galaxy even though it is much smaller; (b) releases approximately 100 times as much energy as a large star; (c) releases more than 100 times as much energy as a large galaxy; (d) is much smaller than a star the size of our Sun.

20. Which is the correct sequence of events that occurred during the evolution of the Universe?
(a) Atoms formed, then planets formed, then stars formed, then matter collected into galaxies. (b) Atoms formed, then stars formed, then planets formed, then matter collected into galaxies. (c) Galaxies formed, then atoms formed, then stars formed, the matter collected into planets. (d) Atoms formed, then galaxies formed, then planets formed, then matter collected into stars. (e) Atoms formed, then galaxies formed, then stars formed, then matter collected into planets.

True or False:

1. One parsec is equal to 3.2 light years or 3.1×10^{13} kilometers.

2. The sunlight we see from Earth comes from the Sun's core.

3. Particles emitted by the sunspots interrupt radio communication and cause northern lights on Earth.

4. The Sun is the largest and most energetic object in the sky.

5. The stronger gravity in more massive stars makes their hydrogen fusion more rapid and intense.

6. All stars lie in the main sequence.

7. The Sun evolved from a cloud of dust and gas in space and is now midway through its mature phase as a main-sequence star.

8. A red giant is hundreds of times smaller than an ordinary star.

9. Five billion years from now, the hydrogen in our Sun's core will be exhausted and the Sun will expand into a red giant.

10. Originally, all the stars in the Universe were composed of nearly pure hydrogen.

11. If you were to shine a flashlight beam, a radar beam, or any kind of radiation at a black hole, the energy would be absorbed.

12. If a star were orbiting a black hole, great masses of gas from the star would be sucked into the black hole, to disappear forever.

13. The Milky Way rotates about its center once every 200 million years, so in the 4.6-billion-year history of the Earth we have completed 23 rotations.

14. The galactic halo and globular clusters are probably remnants of the original proto-galaxy that condensed to form the Milky Way.

15. When we look at close objects we see what is happening now, but when we look at distant objects we see what happened in the past.

16. The matter we see in the Universe may be only 10 percent of the mass of the Universe.

Completion:

1. One ______ is the distance traveled by light in a year, 9.5 trillion kilometers.

2. The Sun is composed primarily of hydrogen and ______.

3. The Sun's visible surface is called the ______.

4. Jets of gas called ______ shoot upward from the chromosphere, looking like flames from a burning log.

5. During a full solar eclipse the ______ appears as a halo around the Sun.

6. The ______ ______ of a star is its brightness as seen from Earth.

7. A graph that plots absolute magnitudes versus temperatures is called a/an ______ diagram.

8. A star originates when a large mass of dust and gas condenses under the influence of gravity to form a/an ______.

9. After hydrogen fusion ceases in a star, the core contracts, and the outer shell undergoes hydrogen fusion and expands to become a/an ______ ______.

10. When a star dies, it will blow a ring of gas called a/an ______ ______ into space.

11. The material remaining after the explosion described in question 10 will contract to become a/an ______ ______.

12. When a large star dies and explodes, a/an ______ is created.

13. A/an ______ _____ is a residual star in which electrons and protons are squeezed together to form neutrons.

14. A small residual star that emits regular, closely spaced electromagnetic signals is a/an______.

15. A/an ______ is a concentration of billions of stars held together by their mutual gravitation.

16. Large clouds of dust and gas, called ______, exist between the stars in a galaxy.

17. A spherical ______ ______ of dust and gas surrounds the Milky Way's galactic disk.

18. A Doppler shift to lower frequency is called a/an ______ ______.

19. A/an ______ is smaller than a galaxy, emits more energy, and usually exhibits a large red shift.

20. The invisible portion of the Universe is called ______ ______.

Answers for Chapter 21

Multiple Choice: 1. e; 2. b; 3. e; 4. b; 5. a; 6. b; 7. c; 8. d; 9. c; 10. e; 11. c; 12. a; 13. c; 14. c; 15. d; 16. d; 17. a; 18. b; 19. c; 20. e

True or False: 1. T; 2. F; 3. T; 4. F; 5. T; 6. F; 7. T; 8. F; 9. T; 10. T; 11. T; 12. T; 13. T; 14. T; 15. T; 16. T

Completion: 1. light-year; 2. helium; 3. photosphere; 4. spicules; 5. corona; 6. apparent magnitude; 7. Hertzsprung-Russell; 8. protostar; 9. red giant; 10. planetary nebula; 11. white dwarf; 12. supernova; 13. neutron star; 14. pulsar; 15. galaxy; 16. nebulae; 17. galactic halo; 18. red shift; 19. quasar; 20. dark matter

CHAPTER 22

Geologic Resources

Discussion

In this chapter we describe the nature and origin of energy and mineral resources. Petroleum, coal, and natural gas are the sources of most of the world's current energy supply. We describe how these fuels originate, and how they are extracted from the Earth. We also discuss the limited nature of fossil fuel reserves, their projected future availability, and environmental problems that occur during their extraction and refining. Availability and use of nuclear fuels are discussed in a separate section.

We discuss formation of metallic ore deposits, methods of extraction, and projected future availability of critical metals.

We have also included an Earth Science and the Environment box on energy strategies for the United States. This box introduces political ramifications of petroleum imports, prospects for increasing petroleum production in the United States, conservation, and use of non-fossil fuel resources.

The savings that have been realized from conservation, and the potential savings, are enormous. For example, during the 1950s and 1960s, consumption of electricity increased at a rate of 8 percent per year. If this rate of increase had continued, a total of 1750 new 1,000-megawatt power plants would have been needed by 1990. The Atomic Energy Commission predicted that 800 to 1200 new nuclear reactors would be built. New hydroelectric dams and coal fired plants were also planned.

These predictions were incorrect. Between 1973 and 1985, in an era that was supposed to mark the height of the period of construction of new power plants, only 173 large plants were completed. As documented in the text, no new nuclear power plants were ordered between 1978 and 1992. What has happened? Throughout all sectors of our society, people are simply using less electricity. If the projections made in 1970 are used as a baseline, then in the year 1986 alone, energy conservation saved the people of the United States $150 billion in fuel costs. In contrast, the use of all alternative energy sources saved only $200 million in avoided fuel costs.

In addition to the box on energy strategies Chapter 22 contains an Earth Science and the Environment box on pollution from metal mining.

Chapter 22

Answers to Discussion Questions

1. Sedimentary rock for reasons discussed in the text.

2. You would find iron, gold, and aluminum on the Moon because the Moon and the Earth are composed of the same elements, but not coal and petroleum because these fuels formed from the remains of organisms.

3. You would first look for signs of past life. If you found such signs, you would look for rocks similar to those bearing fossil fuels on Earth, or for similar structures that might have trapped, or otherwise preserved fossil fuels.

4. No, coal is a nonvolatile solid. It won't flow or evaporate and therefore it doesn't escape.

5. If rates of consumption remain constant, we will deplete petroleum reserves within the next few decades and therefore an energy crisis seems imminent. However, coal can be liquified, conservation can extend the life of existing reserves, and numerous alternative energy sources are available. In his book, The Ultimate Resource (Princeton University Press), economist Julian Simon argues that we should not underestimate human ingenuity. Just before the dawn of the industrial revolution, an energy crisis occurred in Europe when forests were depleted and the price of wood increased. Then, coal was discovered and replaced wood. The most accessible coal reserves were depleted rapidly and a second energy crisis occurred. This crisis was overcome when steam powered pumps were invented to remove water from deep mines and miners were able to exploit deeper reserves. Mr. Simon argues that just as past energy crises were averted by changes in the resource base and improvements in technology, we can expect similar advances to occur in the future.

6. Metal reserves will last longer if we discover new ore bodies, develop technology to mine and refine low-grade ores more cheaply and thus increase known reserves, recycle existing materials, or consume fewer metals. Rapid consumption and a throw-away society lead to more rapid depletion.

7. Some elements are quite similar in chemical behavior to other elements. This is particularly true of those that are close together in the Periodic Table. If two or more similar metals are present, they commonly dissolve in solutions, and then precipitate together. Thus, a single ore deposit may contain two or more valuable metals. For example, silver is chemically similar to lead, and silver is commonly found in galena, the principle ore of lead. In fact, many galena deposits are mined not for lead, but for the silver. Copper, lead, and zinc are

commonly associated in the same mineral deposit. The manganese nodules of the deep sea floor are typically rich in iron, nickel, cobalt, and other similar metals.

8. Elements are never consumed, they are just dispersed so that it becomes expensive, sometimes prohibitively so, to collect them once again. If rates of recycling increase, the life-span of metal reserves can be extended. Fossil fuels represent stored chemical potential energy and when they are burned, this potential is lost. The Second Law of Thermodynamics assures us that this energy cannot be recovered and reused to produce work and heat.

9. The Romans discovered nearly pure deposits of metals at the Earth's surface. The first oil well, dug in 1859, was only 21.2 meters deep. Today, these rich and accessible deposits have been depleted but machinery is available to extract ores and fuels from deep underground and from harsh environments. If major industries were destroyed by a nuclear war, and the human race survived, it might be very difficult to start a mechanized civilization over again. If there were no machines to dig the iron and drill for petroleum, there would be no iron to make the machines and no fuels to power them. At least in the beginning, people would have to resort to mining old dumps and society would become organized more simply.

10. Each student will have a different answer to this question. It is raised to increase awareness of our dependence on our resource base and to provide a springboard for discussion of recycling and conservation.

11. When easily accessible petroleum reserves are depleted, harsh and delicate environments such as the Arctic or offshore continental shelves become more attractive. Both the probability of accidents and the consequences of an accident are more severe in these environments. After conventional petroleum reserves are exhausted, we will probably mine oil shales and convert coal to liquid and gaseous fuels. Significant pollution problems accompany both of these processes. A similar situation occurs with coal. The most concentrated deposits closest to the surface are mined first. As deeper and less concentrated deposits are exploited, more overburden must be removed per ton of coal extracted.

12. The cost of reclaiming mines is added to the cost of the metal or fuel extracted, raising the price these resources. If mines are not reclaimed and streams are polluted and farmland destroyed, then the cost are said to be <u>external costs</u> or <u>externalities</u>. Externalities are generally not so obvious or easy to quantify as direct manufacturing costs, but they are

nevertheless quite real. Silted streams don't provide the quality of recreation that clear ones do, so people must travel farther for their vacations and the local tourist industry suffers. If mine tailings pollute streams, the cost of re-purifying water downstream must be borne by someone. Alternatively, if the water is not purified adequately, then we must consider the price of pollution which may include lost fisheries, increased medical bills, loss of work because of illness, and possible death at an early age. Note that external costs are not always borne by those who benefit most from the mining operation. For example, when coal is mined and electricity is generated in Arizona, local residents bear the external costs, but most of the power is shipped to California. Note also that externalities include only factors that can be measured in monetary terms. Human pain and suffering cannot be valued in this manner and are therefore not accounted in an economic analysis.

Selected Bibliography

The formation of ore deposits:
K. L. Von Damm: Seafloor Hydrothermal Activity: Black Smoker Chemistry and Chimneys. Annual Review of Earth and Planetary Sciences 18 (173), 1990.

John M. Guilbert and Charles F. Park, Jr.: Ore Deposits. San Francisco, W. H. Freeman, 1986. 985 pp.

C. Meyer: Ore Deposits as Guides to Geologic History of the Earth. Annual Review of Earth and Planetary Sciences 16 (147), 1988.

Robert C. Newton: Metamorphic Fluids in the Deep Crust. Annual Review of Earth and Planetary Sciences 19 (385), 1989.

Mineral resource availability:
Phillip H. Abelson: Future Supplies of Energy and Minerals. Science, 231: 657, 1986.

Data on energy supply and demand are form:
Energy Information Administration: Annual Energy Outlook With Projections to 2010. Washington, DC, USGPO, 1991. 125 pp.

Energy Information Administration: Annual Energy Review 1990. Washington, DC, USGPO, 1991. 330 pp.

Chapter 22

Thousands of books have been written recently on conservation and alternative energy sources. Three of our favorites are:

Christopher Flavin and Nicholas Lenssen: Beyond the Petroleum Age. Washington, DC, Worldwatch, 1990. 65 pp.

Scientific American: Energy For Planet Earth. San Francisco, W.H. Freeman, 1991. 161 pp.

Christopher C. Swan: Suncell: Energy, Economy, and Photovoltaics. San Francisco, Sierra Club Books, 1986. 240 pp.

Chapter 22

Geologic Resources

Multiple Choice:

1. Geological resources include
(a) fossil fuels; (b) uranium; (c) metals; (d) sand and gravel; (e) all of these; (f) a and c.

2. Fossil fuels are
(a) renewable; (b) formed from the remains of plants and animals; (c) found mainly in igneous rocks; (d) quarried out of limestone (e) being formed as rapidly as they are being consumed.

3. Mud converts to shale and organic material converts to liquid petroleum
(a) when pressure and temperature are increased due to burial under younger sediment; (b) very quickly in the ocean; (c) when pressure and temperature are decreased due to burial under younger sediment; (d) in the range of 0 to 30° C.

4. The source rock for most oil is
(a) sandstone; (b) dinosaur bone beds; (c) shale; (d) limestone; (e) lake beds.

5. Oil reservoir rocks must be
(a) porous; (b) permeable; (c) impervious; (d) organic shale; (e) a and b.

6. An oil reservoir is most similar to
(a) an underground pool; (b) a potato sliced for chips; (c) an oil-soaked sponge; (d) a lake of oil; (e) none of these.

7. Methane forms when petroleum is heated above _____ degrees Celsius.
(a) 20; (b) 50; (c) 100 to 150; (d) 500; (e) 1000

8. Petroleum will become scarce in a matter of
(a) years; (b) decades; (c) centuries; (d) millennia; (e) one million years.

9. Coal reserves are thought to be sufficient to last
(a) a few years; (b) a few decades; (c) a few centuries; (d) essentially forever, because coal forms faster than we use it.

10. Low-grade oil shales
(a) require more energy to mine and convert the kerogen to petroleum than is generated by burning the oil; (b) require less energy to mine and convert the kerogen to petroleum than is

generated by burning the oil; (c) can be pumped if detergents are added to the reservoir; (d) are currently being used for fuel.

11. Sulfur waste in mine spoils does not contribute to (a) acid rain; (b) acid mine tailings; (c) surface water pollution; (d) groundwater pollution.

12. The major fuel in nuclear power plants is an isotope of (a) Strontium; (b) Uranium; (c) Thorium; (d) plutonium; (e) sulfur.

13. As liquid magma cools and solidifies, (a) high-temperature minerals crystallize first; (b) high-temperature minerals crystallize last; (c) all minerals solidify into a homogeneous mass; (d) all minerals solidify at once.

14. Crystal settling occurs (a) whenever metal-bearing solutions encounter changing conditions that cause precipitation; (b) when magma cools slowly deep underground; (c) when surface streams slow down and deposit sediment; (d) when landlocked lakes dry up; (e) when the oxygen continent of the atmosphere changed.

15. Hydrothermal solutions (a) are corrosive; (b) are hot; (c) dissolve metals; (d) scavenge metals from crustal rocks; (e) all of these.

16. Large, low grade hydrothermal ore deposits are (a) hydrothermal veins; (b) pegmatites; (c) layered mafic; (d) placer; (e) disseminated.

17. Placer ore deposits form (a) whenever metal-bearing solutions encounter changing condition that cause precipitation; (b) when magma cools slowly deep underground; (c) when surface streams slow down and deposit sediment; (d) when landlocked lakes dry up; (e) when the oxygen continent of the atmosphere changed.

18. Nearly 30 percent of North America is underlain with (a) banded-iron formation; (b) marine evaporites; (c) placer deposits; (d) disseminated ore deposits; (e) gold.

19. The most abundant and economically important banded-iron formations formed (a) between 2.6 and 1.9 million years ago; (b) between 2.6 and 1.9 billion years ago; (c) between 20 and 30 billion years ago; (d) because the early atmosphere was rich in molecular oxygen.

20. As ore concentrations decrease, the energy consumption required to mine and process them

(a) decreases; (b) stays the same; (c) rises rapidly; (d) becomes less important.

True or False:

1. Plant matter is composed mainly of carbon, hydrogen, and oxygen.

2. If conditions are favorable, petroleum is forced out of the source rock and migrates to a nearby layer of sandstone or limestone.

3. Crude oil is a gooey, viscous, dark liquid made up of thousands of different chemical compounds.

4. Coal is forming today in some swamps.

5. Imports account for about 10 percent of the petroleum consumed in the United States.

6. Coal can be used directly in conventional automobiles.

7. On the average, more than half of the oil in a reservoir is left behind after a well has "gone dry."

8. All steps in fossil fuel extraction and refining are potential causes of environmental problems.

9. Small quantities of sulfur are present in coal.

10. When oil shale is broken into small pieces and exposed to the atmosphere, it absorbs water and swells.

11. Placer deposits form from crystal settling.

12. Several times during the history of the Earth, shallow seas covered large portions of North America.

13. Bauxite is an example of a hydrothermal deposit.

14. Pure iron is commonly found in the Earth's crust.

15. If sufficient ore is not available, shortages can be alleviated by substitution.

16. More money has been made mining sand and gravel than gold.

Chapter 22: test

Completion:

1. Petroleum, coal, and natural gas are called ______ ______ because they formed from the remains of plants and animals that lived in the geologic past.

2. ______ is a combustible rock composed mainly of carbon.

3. A commercial petroleum ______ forms when oil flows from the shale and concentrates in other rock.

4. A/an ______ ______ is any barrier to the upward migration of oil or gas.

5. Forcing superheated steam into old wells at high pressure is an example of ______ ______.

6. A waxy, solid organic substance that is the precursor of liquid petroleum is called ______.

7. Rich offshore oil reserves exist on ______ ______ in many parts of the world, including the coast of southern California, the Gulf of Mexico, and the North Sea in Europe.

8. Crude petroleum must be ______ to produce gasoline, propane, diesel fuel, motor oil, and chemicals.

9. Modern nuclear power plants use a process called______ ______ to generate heat.

10. ______ is any natural material sufficiently enriched in one or more minerals to be mined profitably.

11. An underground mixture of hot water and dissolved ions is called a/an ______ ______.

12. A/an ______ deposit is formed when a hydrothermal solution flows through a large volume of country rock to form a deposit with low metal concentration.

13. Table salt and borax form in ______ deposits.

14. Layers of iron-rich minerals sandwiched between beds of silica minerals are called ______ ______.

15. Insoluble ions are left behind by weathering to form ______ deposits.

16. Phosphorus and potassium are examples of ______ geologic resources.

Answers for Chapter 22

Multiple Choice: 1. e; 2. b; 3. a; 4. c; 5. e; 6. c; 7. c; 8. b; 9. c; 10. a; 11. a; 12. b; 13. a; 14. b; 15. e; 16. e; 17. c; 18. b; 19. b; 20. c

True or False: 1. T; 2. T; 3. T; 4. T; 5. F; 6. F; 7. T; 8. T; 9. T; 10. T; 11. F; 12. T; 13. F; 14. F; 15. T; 16. T

Completion: 1. fossil fuels; 2. coal; 3. reservoir; 4. oil trap; 5. secondary recovery; 6. kerogen; 7. continental shelves; 8. refined; 9. nuclear fission; 10. ore; 11. hydrothermal solution; 12. disseminated; 13. evaporite; 14. banded-iron formations; 15. residual; 16. nonmetallic